Praise for The Case for Space:

"Paul Hardersen has written a rare and important book about the exploration of space. While there is plenty of grand vision and high technology here, his book also shows how space technology makes very personal contributions to the quality of life of people around the world. This book shows how individuals can be a part of advancing the space frontier no matter where they are, as Paul invites you to join the great adventure."

> **Scott Pace**
> *Executive Vice President, National Space Society*
> *Chairman, Policy Committee*

"This book is long overdue. Hardersen has marshaled an impressive array of brass-tacks facts documenting the enormous benefits that the space program has brought to the American public. Those skeptical that the expenditures this nation has made in pushing the space frontier are being rewarded by ample economic returns will find in this book necessary medicine. Those already advocating space exploration will find here some of the tools they need to pry open the eyes of even the most closed minded. Read this book and then mail it to your congressman. Hardersen's is a message they need to hear on Capital Hill."

> **Dr. Robert Zubrin**
> *Chairman of the Executive Committee of the National Space Society*

"The Case for Space is just that: a compelling, easy-to-read argument for space exploration. Mr. Hardersen takes a technical subject matter and presents it in a manner appealing to both the novice and the expert. Either way, when you finish the book, you have a clearer picture of the potential of space technology."

> **Senator Chuck Grassley**

"This is a well-put-together collection of all the arguments, both practical and philosophical, for the importance of maintaining a vigorous U.S. space effort. It will serve a space advocate well as source material for pro-space arguments, and it may even convince a few of the unconvinced."

> **John M. Logsdon**
> *Director, Space Policy Institute and*
> *Center for International Science and Technology Policy*
> *George Washington University*

THE CASE FOR SPACE

Who Benefits from Explorations of the Last Frontier?

Frontiers in Astronomy and Earth Science, Volume 3

THE CASE FOR SPACE

Who Benefits from Explorations of the Last Frontier?

Paul S. Hardersen

Foreword by Lori Garver

ATL Press, Inc.
Science Publishers

ATL Press, Inc.
Science Publishers
P.O. Box 4563 T Station
Shrewsbury, MA 01545 U.S.A.

Library of Congress Cataloging-in-Publication Data

Hardersen, Paul S.
    The case for space : who benefits from explorations of the last
frontier / Paul S. Hardersen
          p. cm. - - (Frontiers in astronomy and earth science : vol. 3)
    Includes bibliographic references and index.
    ISBN 1-882360-47-8 (cloth : alk. paper), -- ISBN 1-882360-48-6
(softcover : alk. paper)
    1. Outer space--Exploration. I. Title. II. Series.
QB500.262.H37 1997
303.48 ' 3--DC 20   96-38265
                                                    CIP

Printed in Canada on acid-free paper ∞

Foreword

A new political agenda has befallen our nation's capital. It is one that is embraced favorably by many space activists. There are many ardent supporters of our nation's space program who recognize that the government will not (and should not) be the primary provider of resources for opening the space frontier. There is no doubt that private industry and individual interests will play a key role in society's expansion beyond Earth orbit.

But I believe there is a continuing role for government investment. Investment in space R&D must be a cornerstone of any realistic agenda for national renewal. Not only is the need to explore intrinsic to life itself, but on a practical level we know that we explore and develop space because there are real rewards to be reaped -- political, economic and social.

As the new Republican leadership in Congress struggles to balance the budget, the space program is being cut. But the space program is about economic strength -- new companies, new products and new jobs. The existing US industrial work force is critical to our economy. The aerospace industry has been the only remaining business sector in our nations' economy with a positive international trade balance. A robust space program is necessary to offset defense-related job cutbacks. Economic conversion to civil space from defense programs can be an efficient means of managing this difficult problem.

What's more, space exploration offers the nations of the world the chance to leave the history of warfare behind and work together toward a new, peaceful age. Our changing world political situation can benefit from international cooperation in space. The cold-war was symbolized by the race to the Moon. The new world order could well be symbolized by cooperating in the space arena.

There also is new knowledge to be gained -- about new technologies, about medical advances, about climate control and environmental protection -- that can enhance our quality of life.

The potential of space development is perhaps our greatest incentive. The possibilities of solar power satellites for unlimited clean energy, extraterrestrial materials for manufacturing, and pharmaceutical development for medical research could greatly enhance the future of human society.

Already our space program has benefited our lives here on Earth. Communications satellites have literally opened up the world for exchange of information, ideas and currency. Earth monitoring satellites are allowing us to better understand our global environment. When astronauts first went to the

Moon, perhaps the most important discovery was their view of Earth -- which changed our world perspective forever.

We must take some basic steps now in order to take advantage of these important possibilities. Investing in R&D programs such as Reusable Launch Vehicle Technologies will be critical to the opening of the space frontier. Continuing to gain knowledge about the long-term effects of microgravity on humans will be necessary to fully develop space. Also important are the policies we set with regard to doing business in space.

Future-thinking individuals who are interested in the value of space exploration and development can gain enhanced knowledge about these issues and others throughout the following pages. I challenge you to then use this new knowledge to get involved in supporting our space program. By writing this book, Paul Hardersen has done what many space supporters just talk about doing -- he has made a difference in creating a spacefaring civilization.

Lori Garver
Executive Director, National Space Society

Preface

Greetings! By opening this book you have taken the first step on a journey of discovery. Like all journeys, you will be tested in what you think is real and what is possible. More than 99.99 percent of human existence has taken place on our beautiful blue orb known as Earth; but what about the rest of the universe? We are but a small part of a much larger picture. Despite that, very few people care to know much beyond the tangible events that occur in their everyday lives. This book is about why *you* should be concerned about human expansion into space. Your first question might be: what is in it for me? The answer is: plenty! Before we proceed, however, there are a few things we need to discuss.

First, this book is designed just for you. It does not matter what your job is or your position in life. You may work as a plumber, banker, janitor, engineer, cook, or computer programmer. Maybe you stay at home and raise your children. It just does not matter what you do. This book discusses ideas and plans about humanity's future in space that are, largely, very possible. No pie in the sky here! Everything in this book could very well happen within the next 50 years. That is, if people like you start to see the enormous potential of going into space. So far in the Space Age, we really have not reached out very far. We went to the Moon, but we did not stay. Our shuttles and space stations orbit Earth, but we give scant attention to locales beyond the backyard of our home planet. This must change.

What kinds of ideas am I talking about? What do you think about huge satellites poised in space that collect the unlimited bounty of solar energy and send it to Earth for use as electricity? How about space ships landing on asteroids to mine industrial-use metals from those great big hunks of rock? There will also be new satellites orbiting Earth, new bases on the Moon and even the colonization of Mars if people start to realize the enormous wealth we can obtain. This does not even include the wealth of knowledge gained that will be truly staggering.

We will not only focus on the future, but will also look at the past and the present. What have we received from our investment in space? Spin-offs from space technology are found in more places and products than may be realized. Whether it is medical instrumentation, consumer products one can buy in stores, or the hardware that runs computers, spin-offs are closer than you think.

How much money is spent on space projects? You may think it is a lot, but the actual numbers may be surprising. Space is a low priority within the US government. There is a danger that major expansion into space will never occur, unless you convince our politicians otherwise. On the public scene, most people are not very knowledgeable about the wide array of activities that fall under the heading of the "space program." Many people also do not find space activities relevant to their daily lives. This perception breeds apathy and is reflected in the lack of public interest in space exploits today. The body politic in Washington embodies this apathy by viewing space programs in terms of jobs and dollars, instead of the truly revolutionary benefits that can transform and improve the entire human race. This is the challenge of the space community today: to educate the public and politicians to the bounty that awaits us beyond Earth. This book is but one small part of that effort.

Beyond this, I would like you to adjust your view of existence as you read this book. I ask this, because it is very likely that some of you are going through life without realizing that there is more to reality than just planet Earth. People are too occupied with their jobs, families, hobbies, relatives and friends to have much time to think of the universe they live in. However, you need to do exactly that. We live in the suburbs of the Milky Way galaxy. The Milky Way contains billions of stars and it is just one galaxy among billions of galaxies. It is enormous and sometimes overwhelming to think about the great expanse beyond our home planet. A good suggestion is for you to go out to a dark place on a clear night and just look up at the stars. Dark places are specially good, because you can see more stars and even the glowing band of the Milky Way. Get a blanket, lie down on your back, and ponder the wonder of it all. While doing this you just may begin to realize that human society must go outward. Space is just an extension of our environment and nothing is stopping us from going. Whatever thoughts cross your mind during this night, remember them as you read this book.

I hope that you enjoy the wonderful possibilities that I will try to convey to you here. Human society is at a very critical juncture in its existence. Hunger, overpopulation, pollution, and a retreat from science and rational thought all threaten to send our civilizations back to the Dark Ages. Large problems need bold and visionary solutions. Maybe some of the ideas here will become a part of our future that will usher in the Golden Age of Humanity. At the very least, they are possibilities you should consider.

Paul S. Hardersen

Contents

Cover illustration: Map of Earth's seafloor gravity obtained from Geosat
satellite images collected over an 18 months period. The Geosat satellite cost
$80 million. It would take several billion dollars and over 100 years to obtain
an equivalent chart, using shipboard topographic equipment. Areas of high
gravity are in yellow and orange, low in blue and purple. Courtesy Dr. David
Sandwell, Scripps Institution for Oceanography, La Jolla, and Dr. Walter
Smith, National Oceanic and Atmospheric Administration, Washington. The
permission to reproduce this image does not constitute an endorsement of the
views expressed in this book. The image was publishd in Earth magazine
(June 1996), whose editorial office provided kind assistance.

Introduction

In the distant future, historians may record August 5, 1996, as one of the most pivotal turning points in human history. This date may forever be linked to the intellectual, scientific, social and philosophical changes-of-thought that occurred as a result of the discovery of independent life on Mars. Researchers at NASA and Stanford University announced on this day their observation of evidence in a meteorite (found in the Antarctic in 1984) for what is believed to be the presence of bacterial life originating on Mars about 3.6 billion years ago.

The findings, if confirmed, suggest several things. First, this is an additional link in the chain of evidence needed to show that other intelligent life exists in the universe. It is a very long process to go from a simple single-celled organism to a biologically complex creature, such as a human, but this moves us in that direction. Second, this finding corroborates the view that early Mars was warm and had water. Where there is water, there is life -- and this finding may again prove that. Third, the philosophical implications are staggering. Human self-centeredness prevails today, even though Copernicus showed centuries ago that we are not even the center of our own solar system. This discovery is also another step up on the ladder of realization that we are only a small part of the expansive, beautiful universe. Instead of despairing over a sense of insignificance, we should rejoice that we exist on our home world and have the ability now to explore what the rest of the universe has to offer.

The discovery of life on Mars is another compelling reason for humans to increasingly explore our solar system and galaxy. The search for other life -- intelligent or not -- is a goal that will surely excite the public, the space community, and our elected officials. It is time for people to realize that we are not really children of Earth, but children of the universe. The Mars finding may provide the impetus for us to finally commit to space exploration and true discovery. What we will find is indeed be hard to predict!

The universe of space exploration is as varied and expansive as the cosmos itself. Space activities used to be simpler and more focused in the past, but today the term "space exploration" encompasses a wide variety of activities and missions. Some of the work includes research into new, economical launch vehicles. There is the constant gaze of satellite eyes that study the Earth and the cosmos simultaneously. New ideas are being espoused that promote the renewed human exploration of the Moon and expeditions to Mars. Many companies, such as Motorola, are investing heavily into space-based communication systems. The flow of technological benefits from space research to the general populace swells steadily and becomes more apparent over time.

It would be very difficult to write one volume about all the varied elements involved in space-based work. Therefore, the goal of this book is to introduce you to a sampling of some of the vital work in the space community. This volume will hopefully wet your appetite for further research and investigation into whatever space subject interests you. I also hope that this book will reveal many of the benefits and potential benefits, of exploring the rest of the universe. Earth may be our home, but our self-image as a species will be incomplete, if we do not recognize that we are creatures of the entire cosmos.

The six chapters in this book cover seemingly diverse issues. However, the topics are all related. Satellite operations, scientific probes, new launch vehicles, spin-offs and dramatic new goals, such as space solar power and asteroid mining, coalesce into a single vision. That vision has all space activity working simultaneously toward improving everybodies' lives on Earth and exploring a universe that is still relatively new and unknown to us.

Chapter 1 introduces you to the tangible benefits of space flight -- namely, spin-offs and some of their effects. Many people have heard of spin-offs, but few know of their real impact on society. Spin-offs can be viewed in a macroscopic or microscopic manner. The large picture reveals that spin-offs touch most people's lives at one time or another. A closer view portrays some very personal stories that have dramatically changed people's lives for the better. People must connect with their space program and its activities. If we ignore it or fail to realize the connection, public support for space efforts will continually erode.

The next chapter highlights some direct benefits from space missions and a describes a few of the space frontiers that are currently being probed. Remote sensing satellites, just as weather or direct broadcast satellites, have been on the international scene for decades now and are used in ways that people can hardly imagine. The power of viewing Earth in different regions of the electro-magnetic spectrum has allowed us to better manage environmental resources, and allocate land for optimal use. It has been used for everything from potential earthquake prediction to the protection of endangered species..

Conversely, a quick review of a few space science missions should remind readers that the quest for knowledge is just as important as the practical benefits of space flight. Researchers in the astronomy and space science fields are enthusiastic, as new and continuing missions reveal more of the universe to us. By the middle of 1997, robotic probes will begin the renewed investigation of Mars. New space-based astronomy satellites will also be launched. Of the many research areas, the continued search for extra-solar planets is one that can intrigue even the most uninvolved lay person. The debate is still raging about the likelihood of other intelligent life in the universe. Increased astronomical capabilities will allow astronomers to soon find Earth-sized planets and to get a better grasp on the question of life elsewhere in our galaxy.

Chapter 3 moves from deep-space to the corridors of power in Washington. The rather small investment in civil space spending by the US federal government is revealed and placed into context with the rest of the federal budget. Since space has not been a high priority in American political circles since the 1960s, it is even more imperative to uncover the benefits accrued from the investment in space. Besides government spending, the growing field of commercial space is also briefly reviewed. After showing the growth pattern of commercial space firms, several companies are spotlighted to show the diversity of efforts being undertaken. Non-aerospace companies are venturing into the space business, because operations from Earth orbit are lucrative in the fields of communications, television broadcasting and messaging services. This field will only continue to grow in the future.

The unique view from Earth orbit also becomes apparent when discussing the environmental future of our planet. Without space-based monitoring efforts, it would be much more difficult to detect ozone depletion and potential global warming. Besides using satellites to study Earth's atmosphere and our effect upon it, there are a variety of space-based scenarios that could improve our environment. The idea of space-based solar power systems -- to collect solar energy and send it to Earth for electrical use -- has been around for decades, but there are now many variants of this idea that may become feasible. Concern over the pollution caused by the mining of Earth for minerals continues. It may be mitigated by the ability to mine asteroids in orbits that range from near Earth to the main asteroid belt beyond Mars. Several other ideas are presented in Chapter 4. I hope that they will spark your imagination and create other ideas on how to use the solar system to improve the quality of life on Earth.

Another vital area of space research is space transportation. This subject is often taken for granted, but the state of space transportation has not improved significantly since the 1960s. Non-reusable rockets derived from ballistic missiles are still used to launch satellites. Even the space shuttle, although partially reusable, has not achieved its goal of reducing space transportation cost. NASA and the space industry are finally seriously addressing the vital goal of reducing the cost of access to space. Plans are in progress on reusable launch vehicle technology aimed at reducing launch costs by at least a factor of 10. Only by through such measures will the space program become more accessible.

However, the space shuttle has served as a very valuable vehicle for learning more about the intricacies of space vehicle operations, while also providing a test-bed for a variety of scientific research areas. The life sciences are one of the most important areas the shuttle explored. Several missions, discussed in Chapter 5, have conducted in-depth research to learn how the human body reacts to microgravity. We must develop counter measures to the negative biological effects to allow longer trips in space.

The last chapter completes the circle as public opinion polls are examined about the public views on space exploration. Unfortunately, the public is not as supportive of space efforts as some space experts assume. While space activities seem to be accepted on their merits, they fare poorly when people are forced to prioritize among a variety of federal programs. The poll data show that a substantial amount of work must be undertaken to educate the public about the value of space exploits. A variety of space-interest organizations are introduced that the reader can join, if they are interested in becoming involved in advocating a human future in space.

After reading this book, I hope you come away with a realization of the diversity of space efforts that are in progress and a realization of the potential of space efforts for improving the human condition. We all want to make the world a better place. Supporting an expansive and aggressive space program just may be one way by which we can assure increased prosperity for our species today and into the future.

Chapter 1: Space Technology at Home

The universe of space exploration deals with many different sciences and subjects, but one of the most widely known aspects of space activities never leaves the surface of the Earth. I am referring to the spin-offs, a general term for the application of space technologies in commercial or industrial processes that find their way into a vast array of businesses. Spin-offs are an important and beneficial by-product of space exploration efforts. From home appliances and sports gear, to computers and fire fighting equipment, spin-offs pervade society in many ways. They probably had an impact on your life at one time or another.

According to NASA, more than 30,000 spin-offs have been created since the inception of the US space agency in 1958. However, it is impossible to know for sure how much space wizardry has migrated to the private sector, because NASA has no means to document this technology transfer.[1] What NASA has done, and continues to do, is to highlight a variety of space-derived products in an annual publication called *Spinoff*. Many of the spin-offs in this book have their origins in this NASA document.

Before going on, however, it is necessary to assign spin-offs their significant role in advocating a strong and aggressive space program. The most important thing to remember, is that we do not need to go into space solely for spin-offs. After all, people and rockets do not zoom off into the great unknown to build a better vacuum cleaner for your house, or a better tire for your car. Spin-offs are an unanticipated consequence of space travel. Technology transfer is not a pre-planned activity, does not occur magically without any effort. People, either individually or within a company, have a desire to build a new product or improve an existing one. In the search to improve their product, researchers may contact NASA directly or stumble upon some space technology they decide to use. The result is a commercial product that has incorporated some element of space technology and is used on the open market. It can be considered a spin-off. This is an unglamorous process that improves society's efficiency through many small steps instead of by the proverbial "giant leap...."

The main goal of this chapter is to provide pertinent examples of spin-offs that either directly or indirectly affected a number of people in society. The highlighted examples have positively influenced many individuals, families and communities. The persuasiveness of the technology transfer process will become apparent from touching stories of how space technology has empowered people and from spin-offs in people's everyday lives.

Industrial and technically oriented spin-offs are not covered here. Space technology has, however, found wide application in many industries. I will show that spin-offs are relevant and have an impact on home life, personal health, community safety and other critical areas. We often measure the efficacy of programs and products in monetary terms. Yet, intangible benefits, such as improved safety, enhanced personal freedoms and a higher quality of life cannot be ignored.

Fire Prevention

Imagine that you have just finished a hard day at work. It is Friday afternoon, and you climb on your trusty city bus that will take your weary body home. The day has turned out to be a fairly major disaster as you seem to have lost an important financial analysis that was going to be used in a presentation today. Luckily, the meeting was rescheduled so you could find the wayward report. That did not stop your boss from nicely chewing up the left side of your body. Oh well...it's over with and I don't care anymore, you think. You lean back in your seat, stretch lazily, and close your eyes while dreaming of happier places. As the bus chugs through the streets, your thoughts wander to a beach that is replete with beautiful white sand, clear water, an impressive surf, and a drink in each of your hands. Ahhh...paradise! When can I go, you mumble to yourself. Your eyelids then start to get heavy and your body relaxes as you mentally transport yourself to your new home.

You and the other passengers are unaware that catastrophe awaits. An old fuel line in the bus's diesel engine is threatening to spray its flammable contents all over the hot compartment. Neglected due to poor maintenance, the fuel line is weak with corrosion and struggles to hold in the fuel. Unable to do so, a wide crack splits open the line and fuel rushes out of the opening. As you and the other passengers think about your Friday night plans, a liquid bomb forms under the bus. Soon, the fuel bursts into flames. Nobody notices at first, but the newborn fire quickly spreads under the entire length of the bus and moves up toward the unsuspecting passengers. You hear the engine stall and you awaken abruptly. Stuck in the middle of an intersection, finger-like flames begin to probe their way up into the passenger compartment. The material on the bus rapidly succumbs to the flames. Suddenly the bus is engulfed in flames in front of you...and behind you! People yell and scream and try to flee. Bodies press up against the emergency escape doors, but they refuse to open. Flames grow larger and smoke fills the bus. What do I do now? I don't want to die! Your mind screams in panic. You cannot escape. You do not know what to do. Acrid and deadly smoke fills your lungs... you cough and gag while pounding your fist vainly against the windows...your last thoughts are of your beautiful spouse and child as consciousness quickly fades.

This scenario could have happened anywhere in the world. Does the story need to have such a sad ending? The problem is that many mass transit systems use flammable materials in the seat cushions and walls. Besides that, the materials also produce dense smoke that can incapacitate and kill the unfortunate people who have to breathe them. Despite bus safety systems and escape doors, accidents do happen. Sometimes there is nowhere to turn, if a fire breaks out in a crowded vehicle.

During space shuttle development in the 1970s, engineers at the NASA Johnson Space Center in Houston, Texas, were searching for an inflammable foam insulation for use on the shuttle. Fire onboard the shuttle is, of course, extremely dangerous because it can jeopardize the crew's lives and the vehicle's return to Earth. The NASA engineers developed a polymide foam that does not ignite when directly exposed to flames.[2] Instead of igniting and spreading fire, this foam just blackens, chars, and disintegrates.

Relating the confined interiors of buses to that of the space shuttle, the firm Solar Turbines International patented this fire-resistant foam and called it Solimide.® The company produced this foam commercially under a contract from NASA. It finally became possible to reduce the risk of death and injury from fires inside all types of mass transit vehicles. As an initial test, the polymide foam covered the car-end doors of the San Francisco/Oakland Bay Area Rapid Transit (BART) System buses. With this development, a new era was ushered in. We will hopefully see mass transit systems across the US and the world adopt this and other fire-resistant materials that can add another measure of safety for their passengers.

In a similar mass transit system, the subway, concerns had developed that an onboard fire could rapidly spread throughout the vehicle through the wiring system. Documented cases of subway fires spreading along flammable cable insulation and wiring have been reported. There is a clear need to reduce such fire risk. Thanks to a Boston company and research by NASA's Jet Propulsion Laboratory (JPL) in Pasadena, California, this threat has been dramatically reduced.[3] Drawing from its work on solid rocket motors, JPL devised a process that filled up the empty spaces in wire insulation with non-flammable, inorganic materials. By contrast, many organic and synthetic materials are very flammable. Organic materials, which always contain carbon, are many times more likely to burn because of their inherent chemical properties. Examples of some highly flammable items are paper, wood, and cotton clothing.

The method of using inorganic materials in wire insulation -- called bimodal distribution -- caught the attention of the Boston Insulated Wire and Cable (BIW) Company of Boston, Massachusetts. Adapting this new insulation to wires and cables, BIW patented their Lo-Smoke® cable assemblies that have since been sold to mass transit systems all over the world. BIW's product turned out to be very resistant to smoke and flames. This, in turn, increases the safety of those who use public transportation systems.

Besides uses in different transportation areas, flame-retardant material is also needed in situations with an oxygen-rich atmosphere. A fire needs oxygen to burn; take away the oxygen and fire is eliminated. Unfortunately, when NASA's Apollo program began in the mid-1960s with the goal of landing people on the Moon, the cabin atmosphere in the first Apollo capsules contained pure oxygen. Fires burn well in our natural atmosphere that is 21 percent oxygen. They clearly burn much more intensely in 100 percent oxygen. The atmospheric pressure in the Apollo capsule was also higher than the standard pressure on Earth. These ingredients were a prescription for disaster and that is exactly what happened. Tragically, a fire erupted in the Apollo 1 capsule on January 27, 1967. It killed astronauts Virgil I. (Gus) Grissom, Roger Chafee, and Ed White III in a matter of seconds. The search then began for non-flammable materials for use in oxygen-rich environments.

Two fabrics, Durette®[4] and Polybenzimidazole (PBI)[5] , emerged in the 1960s to make spacecraft interiors safer for astronauts. The Monsanto Company of St. Louis, Missouri, developed Durette® as a direct result of the Apollo 1 fire. This material is useful because it does not burn or produce toxic flames. Later, Monsanto sold the rights for this product to Fire Safe Products of St. Louis, Missouri, that produces the fabric for varies uses. For example, hyperbaric chambers are employed to treat patients who suffer from an oxygen deficiency. Maybe a family became ill when their gas heater malfunctioned during the winter and they were poisoned from carbon monoxide. These chambers supply a patient with a higher oxygen concentration to facilitate their recovery. Another use is for divers who suffer from compression sickness after a deep dive because they returned to the water's surface too quickly. The pressure differential from the ocean surface to points far underwater is great. A quick return to the surface can cause a diver's nitrogen to bubble out of the blood, causing intense pain or even death. A diver needs to stay in an oxygen-rich atmosphere for a while to recover from this malady. To make this treatment safe, the chamber material must be inflammable. Due to the work of Fire Safe Products, Durette® has been used to prevent fires in hyperbaric chambers. It has also been used by the US Navy in diving chambers to make the walls, furniture, and clothing less susceptible to fire. At the University of Pennsylvania, the Institute for Environmental Medicine has used this fabric for diving research and oxygen therapy. Even auto racing teams outfit their drivers with Durette® and PBI suits to protect them in accidents and fires.

Complementary to Durette® is PBI, another fire-blocking fiber that the Celanese Corporation developed for NASA and the US Air Force Materials Laboratory in the 1960s. PBI does not burn in air and produces very little smoke or gas at temperatures up to 1040°F (560°C). After the commercialization of PBI in 1980 as an alternative to asbestos, Celanese Corporation pushed its use in clothing for heat- or fire-related jobs and for use in airline seats. Some airlines companies have already decided to use PBI-covered seats. Perhaps they were impressed with some of the PBI tests that showed that PBI-covered seats resisted the spread of fire across the width of a seat for two minutes at 1,900°F (1,038°C).

Fire Fighting

Fire fighting is one of the most hazardous public service jobs in the United States. It is a job in which dedicated public servants risk their lives to protect the people and property of their local community. The fire fighter is counted on by all members of a locality to help them in times of need. Just pick up the phone, dial 911, and the fire team is on its way.

Before 1970, fire fighting suits and equipment could be characterized as being adequate for the job of fire fighters, but not much better than that. The breathing apparatus and air tanks were heavy. The fire suits were hot and restrictive. The result produced fire personnel who tired quickly. This prevented them from doing their job efficiently and expediently. In 1971, many American fire chiefs called for research into better breathing systems to help their rescuers do their job better. In response, NASA joined up with the Fire Technology Division of the National Bureau of Standards to find simpler breathing systems.[6] While examining this problem, NASA found two space technologies -- space suits and rocket motor casings. The fire fighters' old air tanks were heavy and quickly caused fatigue. The Martin Marietta Corporation (now Lockheed Martin) and Structural Composite Industries, Inc., used technology developed to make lightweight rocket motor casings to design and build much lighter air tanks than the old cylinders. Besides the new tanks, they modified the entire breathing system to promote easier use by adapting space suit gear. The final product was significantly improved, including a weight reduction from 60 to 20 pounds, and a face mask that allowed better visibility and fit. The system weight shifted from a person's shoulders to the hips, the whole outfit could be donned and doffed more quickly, and an internal alarm in the face mask warned the fire fighter when the air supply was running out.

The new breathing system was initially tested at NASA's Johnson Space Center. Fire departments in New York, Houston, and Los Angeles then field-tested the new equipment under realistic fire fighting conditions in the mid-1970s. These tests gave a very positive response. "It was a major improvement in fire fighting equipment, no question", according to James Manahan of the New York City Fire Department's Safety Operating Battalion.[7] Manahan served as the project officer for NYFD's participation in the 1974-75 field tests. "The NASA technology definitely made a contribution toward reducing fire fighter fatigue."[8] Comparing the old suits with the NASA suit, Manahan noted the old system "was heavy, bulky, had narrow eye pieces and the weight pulled down on the shoulders. When you wore that thing for 15 minutes, you couldn't wait to get out of it. And this" -- indicating the NASA system -- "is the current system we use, with a smaller, lighter air cylinder, better mask and harness."[9]

Work on fire fighting equipment technology did not stop there, though. In the early 1980s, NASA teamed up with the US Fire Administration to conduct Project FIRES.[10] This effort was aimed at improving the overall effectiveness of fire fighting clothes and equipment. Again, space suit technology provided some solutions in the quest to make fire suits lighter and more fire-resistant. Project FIRES, which stands for Firefighters' Integrated Response Equipment System, tested PBI, Nomex®, and Kevlar®. After selecting the right blend of materials, a new suit emerged that weighed only 12 1/2 pounds. This is very light for body suits. This improvement resulted in a 40 percent weight reduction compared to previous fire suits. Once out of the lab stage, the suit underwent a nine-month testing period in 1982 and 1983 in which 14 municipal fire departments across the US applied the suit.

This collection of examples provides a glimpse of the possibilities of technology transfer. Although spacecraft and probes are initially made with only space travel in mind, secondary uses of these technologies lay in wait to solve other unrelated problems. All that is needed is an inquiring, inventive mind and access to the desired technology . Today, NASA promotes technology transfer to the private sector where feasible. There is no better time than now, to think of how the people of Earth can directly use the machinery of space.

Medical Technology

One of the best-kept secrets is how widespread and instrumental NASA technology has been in advancing the medical equipment industry. From the Mercury program to today's shuttle missions, NASA and doctors from around the US have sought to learn how the microgravity environment affects the human body. The lack of gravity tends to do strange things to human bodies. These effects need to be quantified and corrected so that long-duration missions (such as trips to Mars) will not ravage astronauts' bodies to the point that they cannot function normally once they arrive at their destination. It would be quite embarrassing for the first astronauts to land on Mars and discover that they cannot go outside and look around because their bodies are too weak. Some of the known maladies include calcium and mineral loss in bones, body fluid shifts to the upper torso, a lengthening of the spine, back pain, and the well-known nausea and discomfort encountered during the first few days of a space flight. One major reason for building the international space station is the need to assure the human body can stay healthy during long stays in space. Construction of this orbiting outpost will begin in late 1997.

To monitor the health of astronauts in the past, specific instruments and devices had to be developed for use in the space program. Other technology, totally unrelated to the medical profession, has also played a part in improving the health of average Americans. Whether the subject is pacemakers, monitors, or imaging systems, the folks who took us to the Moon also worked to improve the health of all people.

Heart disease is still the number one killer of Americans today. Although there is ongoing research and people are now much more aware of their diets and exercise habits, the battle against this disease is progressing slowly. However, people in high risk categories for heart disease now have a new ally in their fight. Thry can now not only monitor their heart, but devices can also warn them of pending heart attacks. The time saved by early warnings has the potential to save many lives.

A company in Clark, New Jersey, developed in the early 1980s a portable, computerized heart monitor that a person can take virtually anywhere. This little device allows people to lead normal, unrestricted lives and enables them to constantly keep track of their heart beat.[11] The 14-ounce product of Q-Med, Inc., is designed to sense each heart beat and instantaneously determine whether it is normal. If a heart beat exceeds a given limit set by the patient's doctor, the device alerts the patient to see his or her doctor. The benefit of this warning system is that it may be possible to treat heart conditions before they occurs. This saves not only money, but is a much better option for the patient.

NASA participated in this project when the monitor was being developed. To accurately measure each heart beat, the monitor is connected to several electrodes that are attached to the patient's chest. The electrodes sense the heartbeats and tell the monitor how the heart is performing. If a person is carrying this little Q-Med monitor for months or years, it is necessary for the electrodes to stay on the person's chest and not fall off. The data quality will also be improved, if there is a firm attachment between body and electrode. During its research, Q-Med discovered an electrode that NASA used to monitor the heart beat of shuttle astronauts while they were in orbit. Q-Med then quickly decided to use it and received an exclusive license from NASA to market the product. In August 1984, the US Food and Drug Administration (FDA) issued its approval and cleared the way for the Monitor One heart device to reach the market.

The Synchrony Pacemaker System is another heart device that utilizes a different type of space technology. It was developed by Siemens-Pacesetter, Inc., of Sylmar, California.[12] This system includes an improved pacemaker and a portable unit (the PA APS-II) that can store information on the performance of the pacemaker. A doctor then uses this information to program the pacemaker to fit the patient's needs. The APS-II can communicate with the pacemaker by using bidirectional telemetry, which is the same technology used to communicate between satellites and ground stations. This telemetry allows the doctor to program and adjust the pacemaker without having to cut the patient's chest open to get at the pacemaker. This is another example of reducing medical costs. Surgery is avoided and the patient's quality of life is improved. Widespread use of this pacemaker began in August 1989, after clinical trials involving over 750 implants at more than 90 hospitals. The results of these trials were positive and the FDA issued its approval.

Besides the Synchrony system, there are other instances of the use of telemetry to improve the health and quality of life for people. Encouragement from NASA led the Parker Hannifin Corporation of Cleveland, Ohio, to join up with the Applied Physics Laboratory (APL) at Johns Hopkins University to develop the Programmable Implantable Medication System (PIMS).[13] The purpose of this device is to regularly inject precise doses of medication into a specific organ that needs treatment. For example, suppose a patient suffers from *diabetes mellitus*. Most people have heard about this condition and know what the treatment involves -- insulin shots. In this case, the pancreas does not produce enough of the insulin hormone to adequately regulate the body's blood sugar level. Hence, the traditional method of providing insulin to the body is to self-administer shots several times a day. Most people are not too fond of receiving shots, but in this case, patients would have to learn to tolerate daily injections. Now there is an option for diabetics. The PIMS removes most of the needle work and does not restrict patients to a tedious daily schedule of shots.

APL has created external and internal medical equipment for injecting medication into the body. The firm accomplished this by using its experience in small spacecraft design and the micro-miniaturization of electronics together with the development of Parker Hannifin's "peanut valve". The so-called "peanut valve" proved to be very valuable in this effort. Originally, this valve played a role in the 1976 Viking mission to Mars. Located on the landers on the Martian surface, it helped in an experiment that injected tiny amounts of fluid into Martian soil samples that were subsequently analyzed for signs of life. From Mars to Earth, this little valve has applications that span millions of miles.

Work on the PIMS pump, however, dates back to the 1970s and was difficult to perfect. The first PIMS implantation took place on November 10, 1986, and involved F. Jackson Piotrow at the Johns Hopkins Hospital in Baltimore, Maryland.[14] At least 17 PIMS have been implanted since that time. PIMS is about the size of a hockey puck and is surgically implanted into the abdomen. Telemetry permits programming of the unit. In addition, the insulin supply is refilled four times a year by using a special hypodermic needle. The lithium battery that powers the device operates for more than five years before requiring replacement.

Two other companies, Parker Hannifin and MiniMed Technologies of Sylmar, California, market a small, compact, portable external pump that can continuously inject insulin (or any other medicine) to any part of the body. The MiniMed 504® is about the size of a credit card and weighs only 38 ounces. Again, one of the major benefits of this device is that it frees the diabetic from a restrictive life of regimented shots. This is an intangible advantage because one cannot measure the value of being free to do everything and going anywhere compared to being restricted to a certain area for shots. By being able to live life to the fullest, diabetics can count on a much more satisfying existence. It must be attractive, because more than 12,000 diabetics have switched to the MiniMed 504® since the 1980s.[15]

Let us now switch from medical devices for use in or on the body, to medical equipment that permits a view inside the human body without the need for a scalpel. The key process for this is called digital imaging. NASA has used this process for years, ever since it began sending robotic probes to other planets. This is a computer-driven process that transforms pictures into a series of numbers. Say that a probe circling Neptune took some images of that planet's clouds, and wanted to send those pictures back to Earth. The computer on the probe would change the photo into a series of numbers that represented the various features and colors in the photo. It then transmits these data to a ground station on Earth where they are received, and a computer re-assembles the orderly array of numbers into pictures for everyone to see.

Several companies have taken on the task of using digital imaging to give doctors a new tool to view inside the human body. One company formed solely to transfer this technology to the medical community.[16] Started by two former NASA employees, Dr. Kenneth Castleman and Don Winkler, who worked on digital imaging, this firm created a line of computer work stations that performed quantitative digital image analysis. Examples of the uses of digital imaging include, CAT scanners, diagnostic radiography systems, three-dimensional reconstructive techniques in the field of microscopy, and information from cardiological X-rays that show how heart valves and arteries function.

In a specific case, a California firm has used digital imaging to create a radiography machine that supplies a non-invasive view into the human body. Digirad Corporation of Palo Alto, California, introduced its *System One* radiography machine in 1984 with two significant features that set it apart from similar equipment. First, this machine stores bodily images digitally. This is a departure from other radiography machines that use film to store individual body images. *System One* can store 400 images on a single optical disk about the size of an old vinyl record album. According to Digirad, film-less radiography saves money by eliminating the film and the associated equipment to process it.[17] The other major feature is called energy selective imaging. This enhancement permits an operator to remove the ribs and soft tissue from an image to get an unobstructed view of the organ(s) of interest. This allows doctors to make a quicker, more accurate diagnosis and, in turn, may help the patient in the long run.

Philips Medical Systems International, Best, The Netherlands, has found another medical use for digital imaging. Philip's Digital Cardiac Imaging (DCI) system monitors the progress of balloon angioplasty operations.[18] This operation uses a catheter to enter through a vein and snakes its way to an artery that is clogged with fat. A tiny balloon then inflates to open the artery and increase the blood flow. By using image processing technology developed from NASA remote sensing satellites, the DCI system can show different views of the heart. It can also show the progress of the catheter in a blood vessel with still or live images. The information allows this angioplasty operation to be completed very quickly. According to Philips, the DCI equipment is the most widely used of its kind in the world with over 300 DCI's in operation globally and over 100 used in the US.[19] It should also be noted that balloon angioplasty is a fairly common procedure nowadays. In 1977, 12,000 operations of this kind were performed. That number grew to an estimated 562,000 operations in 1992.[20]

Products for Home and Family

Spin-offs are not always associated with highly complex technology, such as medical equipment. Sometimes space-based products find their way into applications that are not considered to be on the cutting edge of technology. Take shoes as an example. Some brands of athletic footwear are the beneficiary of designs used to comfort astronauts' feet on the Moon. This lunar technology has now made its way back to Earth to benefit many more people.

KangaRoos USA, Inc., St. Louis, Missouri, and Avia Group International, Portland, Oregon (a subsidiary of Reebok), both wanted to produce longer-lasting athletic shoes that reduced the impact forces on the users' feet. Regular shoes used a foam that quickly lost its cushioning ability after repeated use. Sports, such as basketball that involve a lot of jumping wear out shoes very rapidly. KangaRoos, working with NASA and other consultants in a two-year research effort, found that they could apply moon boot technology to their shoes.[21] During the Apollo missions of the late 1960s and early 1970s, astronauts wore boots on the lunar surface made of a special three-dimensional material. These gave their feet extra cushioning and ventilation. KangaRoos adapted this design and modified it into a new cushioning system that reduced the impact forces on feet, while improving the shoe's lateral stability. Called Dynacoil®, this athletic shoe cushioning system helps athletes improve their performance and reduces the risk of foot-related injury and fatigue.

Avia teamed up with space suit expert Alexander L. Gross of Lunar Tech Inc., Aspen, Colorado. This seemed to be their key to success in producing their athletic shoe sole.[22] Gross had extensive experience in designing space suits and received NASA's prestigious Apollo Achievement Award for his work. He helped Avia to develop a rigid and flexible midsole based on space suit technology. The key concept used involves bellows. A bellow works by simultaneously expanding and contracting every time a joint is bent. Avia introduced its Compression Chamber midsole that uses horizontal rows and vertical columns of bellows to improve the shoe's cushioning and stability. In addition,

Avia used a stress-free "blow molding" technique to make a one-piece midsole that is more durable than other midsoles. This is the same method for manufacturing the very impact-resistant Apollo helmet and visor used on the Moon.

Another personal item in the home influenced by space technology is something some people cannot see without: eyeglasses or sunglasses. Unknown to most people, glasses have become much safer in the last few decades thanks to some cosmic innovations. Before 1972, glasses were made of glass. Glass has the advantage of being very scratch-resistant, but it also breaks easily and is a safety hazard. To alleviate this danger, the FDA mandated that all sunglasses and prescription glasses had to be shatter-resistant. The search then began to focus on how to make plastic lenses that were as scratch-resistant as glass.

While studying plastic lenses, eyeglass manufacturers learned about the advantages and disadvantages of plastic. One discovery was that plastic lenses could be scratched very easily. This could prove not only distracting to the wearer, but could also block part of the field of vision. Looking for a solution to this problem, the Foster Grant Corporation of Leominster, Massachusetts, borrowed some NASA technology to vastly improve the scratch-resistant qualities of plastic lenses.[23] In the past, NASA had used a highly abrasion-resistant coating that protects the plastic surfaces of aerospace equipment. Foster Grant decided that this coating would work well on their glasses. They used it as a protective layer on the plastic lenses of their Space Tech line of glasses. Introduced in 1984, the Space Tech glasses proved to be five times more scratch-resistant than those of their contemporary competitors.

One of the main benefits of plastic sunglasses is their ability to block ultraviolet radiation from reaching a person's eyes. Ultraviolet radiation is a known contributor to skin cancer and can also cause eye problems. According to Doctor Richard Young of the Department of Anatomy at the University of California Los Angeles (UCLA), "cataract and senile macular degeneration are the two principal causes of vision loss in western nations. A simple, safe and inexpensive way to delay the onset or progression of these diseases is known: wear protective filters that absorb the blue, violet and ultraviolet (UV) when exposure to bright light is necessary."[24] Today it is normal for sunglass manufacturers to advertise this quality as an important feature of their product.

Another entrepreneurial effort, spearheaded by some NASA employees and aerospace contractors, introduced new lines of UV-blocking sunglasses based on space technology. The genesis of this project started with welders in mind. In welding operations workers have to look at the intense flame of the welding arc. Besides being bright, the arc also emits blue and ultraviolet radiation. Continued exposure to this type of environment could cause eye damage and increase the operator's risk of getting cancer. An effort was mounted to design a welding curtain that would block out all the harmful radiation and reduce the health risks. A new curtain eventually resulted from this effort, but that is not the end of the story. Once they accomplished their goal, NASA employees

Laurie Johansen, Paul Diffendaffer, and Charles Youngberg, along with NASA JPL contractor Doctor Michael Hyson, decided to use their new device to market a line of sunglasses.[25] These ingenious people soon formed the Suntiger Corporation. Using the protective method derived from the welding curtain research, the Suntiger Corporation introduced its Avian Orange and PST (Polarized Selective Transmission) sunglasses. Suntiger offers not only sunglasses, but also visors, ski goggles, prescription eye glasses, and industrial eyewear. The Suntiger technology eliminates more than 99 percent of all harmful light, promotes better vision with clearer and sharper images, and shields eyes from very bright lights.

Technology transfer from NASA to the private sector and the consumer is truly an unheralded and prolific feat in modern economics. The previous examples show that the use of space technology for commercial purposes increases the value and return-on-investment of civil space spending. The great benefit of spin-offs involves the barely perceptible increases in the efficiency of people, processes and machines. Spin-offs impact many people at some time in their lives, but the effect is so subtle and so quiet, that few notice it. By taking the time to learn about this activity, more people will realize how important technology transfer is to the quality of their lives.

Economics, though, is only a part of the spin-off story. Scattered across the landscape of America are people who have been personally touched by different applications of space technology. Some individuals have had their lives so profoundly affected by a spin-off that it has become a permanent part of their lives. Other people have been empowered by a spin-off that has allowed them to enjoy happier and more fulfilling lives. Although spin-offs may have some monetary value, the personal value to some recipients is priceless. A few significant stories follow that show the rather tremendous personal impact a wise application of technology can have on a person. One individual, John Zipay, owes his life to a device that was used at the dawn of the Space Age. Amputees James Carden and Amie Bradley are now free to enjoy their favorite activities due to spin-off derived innovations in prosthetics. Stevie Roper, now a teenager living in Virginia, is leading a normal life due to a unique application of space suit technology.

The John Zipay Story[26]

My name is John J. Zipay. I am a NASA Structural Design Engineer who works at the Johnson Space Center in Houston, Texas. The space program saved my life.

I grew up in New York City. I lived and went to school in New York City until I began working at the Johnson Space Center as a cooperative education student in 1984. Upon graduating college in 1988, I began working full-time at Johnson Space Center.

I could write about my love of the space program as a child. I grew up watching the final Apollo missions, Apollo 16 and 17. I watched Skylab launch into space and come hurtling back to Earth. I watched the two Voyager spacecraft pass Jupiter and Saturn, as well as the early Space Shuttle flights, with the dream of working for NASA when I grew up. With the support of my parents and much hard work, I have a wonderful career. In that sense, the space program is my life. All the wonderful things I have experienced in my life, I owe to the space program.

However, this book is about space technology, its spin-offs and the tangible, concrete and dramatic ways the space program affects people's lives. On November 7, 1991, technology derived from the space program saved my life.

Until I moved to Texas in 1988, I suffered from the minor allergies of pollen, cats, and dogs that tens of millions of Americans deal with every day. Over-the-counter antihistamines controlled my occasional sniffles. However, upon moving to Texas, with its wide variety of grasses, trees and molds, my allergies grew progressively worse. Twice in 1989, I suffered the first stages of anaphy-lactic shock. I twice experienced the early pases of a violent allergic reaction that can lead to respiratory failure, cardiac arrest and death. The reactions were traced to some common molds present in the grasses and wet dirt that I had played football and Frisbee in on those two occasions.

Fortunately, a hospital was close by and the emergency room treatment was sufficient to end the reactions each time they happened. After consultation with an allergist, I had a prescription for a device called an "EpiPen." This device is a cylindrical spring-loaded injector about six inches long and a half-inch in diameter containing 0.3 milligrams of Epinephrine. If I was to have a more severe anaphylactic episode, I could inject the "EpiPen" into my thigh, through my clothes, if necessary, because the spring provides enough force to penetrate the material. It is manufactured by Survival Technologies Incorporated of Bethesda, Maryland, and it is an outgrowth of the Mercury program. The Mercury program placed the first American, Alan Shepard, into space on May 5, 1961.

When people first ventured into space during Project Mercury, the well-meaning medical community expressed grave concern about the effects of weightlessness on the human body. We take for granted weightlessness now as we see the space shuttle crews on TV happily floating in the crew compartment. However, when we took our first steps into the unknown, the medical community was very concerned that humans might not be able to see because the muscles under the eyes would not be needed to support the eyes in a microgravity environment. They worried that digestion and circulation might not take place as they do on Earth. They also feared that the fluids in the inner ear, without the force of gravity to hold them down, might move randomly in space. This could disorient an astronaut to the point that he or she might be too sick to function.

To counteract possible motion sickness, the Mercury astronauts carried a spring-loaded injector filled with anti-motion sickness medication. If a Mercury astronaut, such as John Glenn, became disoriented, he could grasp the injector with his gloved hand, place it against his thigh and push hard. The spring loaded injector would be powerful enough to puncture his heavy space suit and provide an intramuscular injection of the medication he needed.

This injector was never used on a Mercury flight because, as we know today, the effects of weightlessness are not as violent as originally thought. However, the technology developed for the Mercury program was passed along to millions of people, such as myself who may need to inject themselves with medication in an emergency. In the case of my allergic reactions, the medication is Epinephrine. Epinephrine requires intramuscular injection; that is why the injection site is the thigh rather than the buttock. Epinephrine was powerful enough to forestall a very violent reaction so that I could be rushed to a hospital. That violent reaction occurred during the early morning hours of November 7, 1991.

After the two episodes in 1989, I did not have another reaction for over two years. On October 26, 1991, I threw a Halloween party at my house in Friendswood, Texas. Because I am an engineer and a science fiction fan, my friend Peter Fantasia and I turned my home into a "spook house" with a science fiction theme. We had eggs from the movie "Alien" that we had built, a "Star Trek" transporter room, a spacecraft landing bay in the garage, and other outlandish creatures throughout the house. We attracted hundreds of people and a local radio station to the party that evening. Because of the complexity of the party, I did not get a chance to shampoo the rug immediately after the party; there was simply too much to clean up! That oversight almost cost me my life.

On the evening of November 6, 1991, I was lying on the living room rug doing some homework. Evidently, the rug I was lying on contained a considerable amount of the mold allergen. The hundreds of people had carried it in who had walked through my house from the outside to see the displays.

At 2:15 a.m. on the morning of November 7, 1991, I awoke from a sound sleep. I had the sick feeling of diarrhea. I thought it was merely going to be one of those upset stomach nights. However, 30 seconds later as I sat on the toilet, I knew something else was terribly wrong.

Anyone who has had an anaphylactic reaction knows the incredible speed that it overtakes every aspect of your body's functions. The total upheaval of your system starts in under a minute and, if left unchecked, will end in death a few minutes later. I will describe it to give others who have not experienced it some understanding. If they are ever a party to it, they will then know what is happening and may be able to get help.

I threw up a few seconds after I got off the toilet. Red hives started to appear all over my body. My crotch and scalp began to violently itch and, as the hives spread, my entire body started to itch. My face began to swell up and my eyes began to swell shut. My bronchi began to constrict and I wheezed much like an asthmatic. I was dying.

Without delay, I reached for the "Epipen", removed the gray safety cap and pushed the tip into my thigh. The design of the injector had to be simple for an astronaut to use with a gloved hand in outer space. This simplicity of design allowed me, a frightened and dying man without medical training, to quickly inject life-saving medication into my body.

The Epinephrine took effect almost instantly. The reaction subsided to the point where I could dial 911 and have the Clear Lake Emergency Medical Service rush me to the nearby Humana Hospital for treatment. The "Epipen" injector does not cure this reaction. It is a stop-gap measure that kept me alive long enough to reach a hospital for treatment. It was another seven hours before the medication provided to me at the hospital ended the terrifying episode.

Needless to say, when I came home I shampooed the rug and bought myself another injector. I carry the "Epipen" with me everywhere I go. I have had a minor episode in the year that followed my brush with death. I have not had to use the injector since that early morning in November 1991, but I know it will work if I have to use it.

I am alive today because money was spent developing simple, reliable technology for humans to use in outer space. That technology is just as useful and reliable on Earth. With the millions of Americans who suffer from anaphylactic episodes, I know my story is not unique. I hope the reader will look around and see the cordless tools, the calculators, home computers, cellular telephones, microwave ovens, Teflon pots, and satellite television shows to appreciate what the space program has done for our lives. I hope they will also realize that the medical benefits emanating from the space program do save lives. We must continue to explore space. The new questions that we answer, the new problems that we solve, and the new frontiers that we conquer when humans explore space provide benefits that are unforeseen and important to every human being on Earth. We must "boldly go where no one has gone before" because we will be the better for it.

People Helping People

I received the next two stories from the NASA Marshall Space Flight Center in Huntsville, Alabama, in response to a request for information for this book. The two stories vividly describe how ingenuity, compassion, and technology can come together to help people live fuller and happier lives. Try to look beyond the details of these press releases to see the human side of the story. They show how Doug Kennedy is continuing his quest to be a top wheelchair racer in the world. James Carden can now accomplish the normal activities of life even with the absence of his left hand.

Paralyzed Wheelchair Racer

A few years ago, a healthy man from Haleyville, Alabama, Douglas Kennedy, fell from the bed of a truck, and was paralyzed below his rib cage. Kennedy has since become a world-class wheelchair racer with many world records and a gold medal from the 1990 Goodwill Games in Seattle, Washington. Today, thanks to a group of engineers experienced in human-powered vehicle design and construction, and with the support of NASA and NASA contractors; Kennedy competed in the World Para-Olympics for physically challenged individuals in Barcelona, Spain, two weeks after the World Olympics were held there.

Kennedy also hopes to use the world's first composite racing wheelchair in a number of wheelchair races conducted each year in the United States and overseas. These events give physically challenged athletes, such as Kennedy a chance to compete against other physically challenged individuals from around the world. This enriches their lives and provides opportunities to demonstrate their ability to cope with their physical limitations.

Kennedy is the holder of three world records (5 kilometer, 10 kilometer, and 5-mile race) and is the national marathon champion.

Kennedy and many others in the world in his situations are trailblazers, opening a specialized range of athletic endeavors to millions of physically challenged individuals.

NASA's Marshall Space Flight Center in Huntsville, Alabama, became involved with wheelchairs after the Center's Technology Utilization Office received an Easter Seals Association request for assistance. The association asked whether the Marshall Center had anyone with the expertise to develop a flexible wheel for wheelchairs to allow physically challenged individuals to negotiate curbs and other obstacles.

The Technology Utilization Office sent the request to several branches of the Marshall Center. William Snoddy, the Center's Deputy Director of Program Development, saw the problem statement and suggested the Technology Utilization Office contact Kennedy to get a wheelchair user's point of view. When asked by Marshall Center engineers to lend his expertise to the design of a flexible wheel for a wheelchair, Kennedy agreed.

The Marshall Center felt justified in pursuing this request, seeing it as a valid biomedical application of NASA- and NASA-contractor-derived technologies. Technology transfer's ultimate goal, according to Technology Utilization Office Director Ismail Akbay, is to help American industries, educational institutions and individuals stay at the forefront of global techno-logical leadership, including the field of biomedical appliances, such as wheelchairs.

During one of Kennedy's first visits to the Marshall Center he went on a tour of its facilities, including the new Productivity Enhancement Complex. There he

Douglas Kennedy shows how light his new racing wheelchair is by holding it over his head. (Courtesy Adeline Byford, NASA)

learned of work underway to find new applications for composite materials. He also learned epoxy resin/graphite compounds were, pound for pound, stiffer and stronger than steel.

Kennedy asked if epoxy resin/graphite could be used to build a racing wheelchair frame. John Cranston, a Martin Marietta contractor working for the Marshall Center and an expert in making human-powered racing vehicles from composite materials, asserted it could. He added that lightweight titanium could replace the heavy steel portions of Kennedy's wheelchair.

Working in their private time, the design team of Cranston, Morgan Andriulli of Boeing Huntsville, Fritz Gant of the University of Alabama in Huntsville, Alex von Spakovsky of the US Army Missile Command's Structures Division, and Jeff Linder of the Marshall Center designed a racing wheelchair for Kennedy.

As the program developed, Marshall Center engineers John Vickers and William McMahon, along with machinist-technician William Norton of the Center's Productivity Enhancement Complex, added their talents to those of Cranston and his design team. Also volunteering their time and skills were

Martin Marietta contract employees Larry Pelham, Robert Carter, and Phillip Thompson; and Ver-Val contract employees Charles Anders, Craig Wood, Wyatt Poe and George Rittenhouse.

The NASA and contractor engineers and technicians were able to fabricate a "T"-frame racing wheelchair using graphite/epoxy resin and titanium. They used materials left over from a Space Shuttle payload developed at the Productivity Enhancement Complex

Cranston served as project engineer and Carroll handled composite machining. Kirch worked on composite fabrication and machining; Norton machined the aluminum fittings; McMahon coordinated the work of the various contractors and handled acquisition of the necessary materials. Pelham designed the filament winding pattern. von Spakovsky performed the structural analysis. Pelham, Carter and Thompson handled the graphite/epoxy resin filament winding of the "T"-frame tubes. Anders, Wood, Poe and Rittenhouse looked after the titanium welding assembly; and Rittenhouse also managed the plating process.

Several Marshall Center employees also participated in the program: Tenina Bili worked on metallurgical identification; Mike Longmeyer and Jerry Oakley handled load control; Neil Tyson served as test engineer; Vandel Hall and Greg Tanner worked on test fixtures and setup; Ver-Val's Richard Messervy was in charge of test fixture and setup and foam machinist Dee Marlin of USBI.

The new chair is four times stiffer than Kennedy's present design. Stiffness is a key factor in improving the wheelchair's performance. When Kennedy thrusts the wheels of his present chair forward, the frame bends from 1/8th to 1/4th of an inch, particularly in areas with several structural welds. According to the design team, the "T"-frame's stiffness should give Kennedy a significant competitive edge.

The work at the Marshall Center will benefit more than Kennedy and a potential handful of other wheelchair racers in the United States. A conventional wheelchair incorporating epoxy resin/graphite compounds and titanium could be of great benefit to anyone whose age or infirmity make it difficult for them to manage today's heavier designs.

Also, improved devices, such as walkers, braces, splints, frames for invalids' Stryker beds, stretchers, Stokes rescue baskets, portable intravenous bag poles, crutch tops, elevated toilet seats, and canes of epoxy resin/graphite compounds could make it easier for millions everywhere to cope with age or physical disabilities.

The innovative design concepts that resulted from the application of joint NASA, Army and NASA-contractor technologies to this problem led to several patent applications. The patented technology is available for a royalty to wheelchair manufacturers for incorporation into their products. Once manufacturers accept the new materials and inroads are made into the industry, there is virtually no limit to the materials' applications.

Anyone desiring more information on the use of graphite/epoxy resin compounds in the biomedical field may write the Marshall Center Office of Technology Utilization, Mail Code AT01, Marshall Space Flight Center, Alabama 35812, or call (205)-544-2223.

Marshall Center Helping Amputees

A team of NASA engineers, technicians, and support personnel at the Marshall Space Flight Center in Huntsville, Alabama, a representative of one of the Center's contractors, and a certified prosthetist orthotist (CPO) in Huntsville, have pooled their talents and resources to help individuals who have lost a hand.

Hearing that James Carden, a retired Marshall Center aerospace engineer, had lost his left hand in a planer mill in 1986 while working at his lumber business, a team of his former co-workers and friends at the Marshall Center came together to offer their help. Carden accepted.

The available conventional cable-operated hook prosthesis was not strong enough and was too cumbersome to allow him to lift lumber or enjoy his favorite pastime, fishing. He wanted a heavy-duty device to which various attachments could be connected to enable him to perform a broader range of activities. Unfortunately, no one manufactured such a device.

Carden's long-time friend and co-worker, Jewell G. "Pete" Belcher Jr. of the Tethered Satellite System Project Chief Engineers Office at the Marshall Center was a team member that helped Carden. According to Belcher, "studies have shown that the work we proposed to do for Jim could benefit many thousands of other Americans, as well as uncounted others around the world. It is difficult for many of these amputees to perform tasks, such as gardening, hunting and fishing, putting their hair up in curlers, use chain saws, and innumerable other activities requiring two hands."

The volunteer team originally included several of Carden's former co-workers: Belcher; Marshall Center materials engineering technician Willie Norton, who was a co-inventor of many of the team's innovations; Tommy Vest, an engineering designer with Management Services Inc., a branch of Bionetics, Inc.; and John Richardson, manager of the Marshall Center's Office of Technology Utilization. Later, Marvin Fourroux, CPO, of Fourroux Orthotics and Prosthetics, Inc., in Huntsville, joined the team to offer his professional guidance. The team made sockets and suspension systems for both amputees on which the various modalities were tested and evaluated.

In its charter, NASA is charged with facilitating the transfer of its technology to the benefit of American businesses, industries and individuals. NASA Headquarters authorized the Center's work on the improved prosthesis under this clause of the agency's charter. NASA considered the work as a biomedical applications project in its overall technology transfer program.

James Carden uses a NASA prosthesis to move planks. (Courtesy Emmett L. Given, NASA)

Once the project was well underway, the team decided that women as well as men could benefit from its efforts. Richardson interviewed Amie Bradley of nearby Arab, Alabama, who had lost her left hand in an automobile accident. She volunteered to participate, along with Carden, in developing, testing and evaluating various new attachments for their prostheses. As a start, both individuals were equipped with a new, specially modified socket designed to accept the various attachments (or modalities as they are known in the prosthetics industry). As a Wal-Mart employee, one of Bradley's new modalities enables her to work more effectively for her employer.

Working with each individual's prosthetic socket, the team developed a wide variety of modalities. Carden, for example, wanted a heavy-duty lifting

Amie Bradley uses a NASA prosthesis to hold a pot. (Courtesy Emmett L. Given, NASA)

attachment to help him in his lumber business, a chain saw adapter, and a fishing reel crank device. Carden asked if the team could develop a device to hold carpentry nails. The team accomplished that by developing a magnetized attachment.

Both individuals could use a clip attachment for holding small objects. Carden wanted to hold fishing lures while he attached his fishing line to them, and Bradley needed it for items, such as a nail polish applicator. She could also use a hair dryer holder, a hair curler holder, a bowl and pot/pan gripper for use in the kitchen, a spatula-like lifting device for pots and pans, and a carving fork attachment. Both also agreed on the need for an attachment to hold broom and rake handles to make sweeping floors and raking leaves easier.

The fishing reel crank adapter is Carden's favorite. "Now I can still do the things I want to do and smile about it", he observed.

Fourroux found that most new amputees have a general feeling of frustration with their prosthesis at first. This is the first hurdle amputees have to overcome.

Stressing the need for proper occupational therapy, Fourroux continued, "Most amputees are restored to a certain level of function and eventually use their device daily. The modalities are a way of making the prosthetic device easier to use. They are better tailored to specific individual life styles." Used in conjunction with occupational therapy, these NASA-funded devices open new horizons of activities for amputees.

The team at the Marshall Center is working now with a rehabilitation facility to mass produce several of the new devices they have developed for amputees elsewhere in the world. As an expert on prosthetic devices, Fourroux feels those already developed have a range of potential applications and will soon provide amputees with many previously unavailable options for becoming more independent.

"I enjoy working with the Marshall team. The devices that come out of our effort will prove invaluable, particularly to the working amputee. I'm looking forward to my next opportunity to work with NASA", he concluded.

The Technology Utilization Office at the Marshall Center is actively seeking firms that are interested in the commercial manufacture of the new modalities. For the modalities patented jointly by NASA and the Marshall Center inventor, a small royalty is charged. Richardson cautioned that not all the modalities are patentable.

Anyone needing more information regarding the prosthesis work underway at the Marshall Center should contact Richardson in the Technology Utilization Office at (205)-544-0964.

Helping Children

Once in a great while, a person is fortunate enough to be touched by another human in such a way that the effect is felt throughout a lifetime. Amidst all the conflict, hatred and self-interest gripping our society, a sincere act of kindness can restore the hope that people do have the compassion and desire to help each other. This is a story that includes not one, but many acts of kindness that began in 1987 and continues today.

Let us travel to Waynesville, North Carolina, and make a stop at the home of Stevie Roper. As any young boy, Stevie had dreams of taking on the world. Maybe he thought of being a race car driver, a construction worker, a football player, or even an astronaut. He wanted to play with his friends and do all the things that kids want to do, but he was unable to do so.

Since birth, Stevie suffered from hypohidrotic ectodermal dysplasia (HED). HED is a condition when a person is born without any sweat glands. Sweating is the primary way for us to dissipate heat from our bodies. Without these glands, serious problems can arise. This is not a fatal condition, but it restricts a person's activities. Since the body cannot cool itself, a potentially deadly overheating can occur. Consequently, Stevie's early childhood left him unable to play baseball, ride his bike, or even to venture outside on a hot day. His body had to be kept cool. That was more important than the pleasures of childhood. Sometimes while in school, Stevie had to dash off to the boy's locker room and shower fully clothed, because he knew he was overheating. Stevie's situation, accompanied by the normal difficulties a young boy experiences, was exacerbated by his shyness. It also did not help that his classmates avoided him, only because they did not know what was wrong with him. School can be tough sometimes.

Local parents raised concerns about Stevie's presence in school, because they did not know what was wrong with him. They reacted to unsubstantiated fears, in a way similar to that of school children with the AIDS virus. As a result of this ignorance, some parents told their children not to play with Stevie. For any child, this kind of shunning would do nothing to build self-esteem. It made Stevie talk to only very few people. However, his future was to be touched by NASA, space suit technology, and an aunt who wanted to help Stevie lead a fulfilling and unrestricted life.

A car ride with Stevie and his sister turned out to be the catalyst that sparked Sara Ann "Tootsie" Moody, Stevie's aunt, into action. Although precautions were taken to keep Stevie cool, the lack of air conditioning in the car and the hot weather soon began to take their toll on the youngster. Noticing that Stevie was reacting to the heat, his sister quickly pulled over to the side of a road and commandeered a garden hose to cool Stevie off. This traumatic event prompted Sara Moody to begin a search for ways to help Stevie. During her journey of compassion, Sara has not only helped Stevie, but also hundreds of children around the world who suffer from heat-related disorders.

The first thing Sara did after Stevie's close call, was to contact the NASA Langley Research Center in Hampton, Virginia, to see if there was any way that NASA could help her nephew. After some inquires, NASA referred Sara to Life Support Systems Incorporated (LSSI). This company uses space suit technology to create personal cooling systems for people who work in a wide variety of hot environments. LSSI, which markets its Micro Climate Cooling System, adapted NASA's space suit undergarment for its uses. The undergarment, which is essentially a body suit with tubing attached throughout, circulates a chilled fluid that keeps astronauts cool while working outside a spacecraft.

However, the first obstacle Sara faced dealt with money. After discussing the situation with Sara, LSSI decided that it would be able to tailor a suit to Stevie's needs. The bad news was the $4,000 price tag. Even though LSSI later brought the suit cost down to $2,600, Sara did not have the money to buy the unit. Down, but not out, Sara began to canvass her local community for donations. Her first big break came when the Po Folks restaurant chain in Virginia agreed to allow Sara to place large collection jars in some of the restaurants to raise money for helping Stevie. Thus, many people became aware of Stevie's condition and made a small contribution. Tom Arney, an executive from Chesapeake, Virginia, also initiated another fund-raising effort. Within only five weeks, these campaigns raised an astounding $5,000.

Soon the time came to give Stevie what could be truly called a gift of love. It could also be called a gift of freedom. He received it in October of 1987. Stevie's suit turned out to be a miniature version of Life Support's Mark VII Micro Climate System.® It included a vest and a headcap within which an ice-cold liquid circulated. The system's design eliminated 40-60 percent of Stevie's body heat and lowered his heart rate by 50-80 beats per minute. After an initial trial and adaptation period, Sara and others noticed that Stevie's personality was blossoming. He shed the shell of doubt that he had hidden under for so long. According to Sara, "he became a different person. No longer withdrawn, no longer frail but filling out -- it's as if he'd been in a closet, afraid to come out, but now he's very happy. The cool suit has changed his whole life. He has lots of friends now and kids call him up and ask him to play."[27] If Stevie had been the only child to receive a cool suit, then this surely would have concluded as a very happy story. However, Sara had a larger vision of helping many children who were limited by events that were beyond their control.

After Stevie received his suit, Sara began to receive phone calls from people who had heard about Stevie's story and the technology that changed his life. The calls came in from everywhere and included locales, such as India, New Zealand, Kuwait, Saudi Arabia, and Argentina. Recognizing the need, Sara founded the HED Foundation with the goal of providing cool suits to all the children that needed them. "When Stevie was born, doctors thought there were only about 20 HED cases in the United States", commented Sara. "It is now estimated that there are about 17,000 HED cases in the US." Besides HED, many other diseases cause heat-related problems in their victims. Some of these diseases include lamellar ichthyosis, Cursayer Syndrome, cystic fibrosis, and multiple sclerosis. Severe burn victims also frequently develop problems with the heat control mechanisms of their body.

As of May 1996, Sara and her pioneering foundation have distributed more than 260 cool suits. Each child has a special story to tell and it is unfortunate that they cannot all be mentioned here. However, it is appropriate to give a few examples to provide a better understanding of the impact that these cool suits have had on these children.

In the case of Krystal "Toby" Sharret, she lacked sweat glands only in her feet. Although a localized condition in Toby's case, this was still a problem. One day, Toby's mother happened to watch a NASA documentary detailing Stevie Roper's story and she wondered if the cooling of Toby's feet would stop her sores from forming. Toby's mother then tried to freeze her daughter's shoes in an attempt to make them cool before they were to be worn. After doing this, Toby's feet showed an improvement. Sara soon received an impassioned phone call from Toby's mother. As she burst into tears while talking about the NASA show, she explained to Sara that "the tears she was crying were not of pain but of joy." Time had become very critical as Toby's feet were scheduled for amputation. Until then, there seemed to be no solution for Toby. Nothing else had cured her feet and the doctors did not know what else to do. Thanks to the foot panels that Toby received through the HED Foundation, her feet cleared up. She began to participate in some of those normal teenage rituals, such as going to high school and attending the prom.

Another example involves Turk Odom, a young man who had a terminal case of epidermolysis bullosa. Although Turk died of his condition near his 18th birthday, the use of a cool suit from Moody added considerable quality to his shortened life. Before receiving his cool suit, Turk's medical condition could only be described as unbearable. Blisters covered his body. In a two-year period, Turk could not eat or drink because his esophagus had closed up. He therefore needed a feeding tube that went directly into his stomach. The pain he endured was continuous and ever-present. At the age of 14, Turk received his cool suit on the night of February 14, 1988. Turk's turnaround astounded everyone. After wearing the suit for six weeks, Turk gained weight at the rate of about one pound a week. His constant pain also subsided. With these changes, the teenager became more lively and his spirits were buoyed. He even drank apple juice. This was no small feat as Sara said that, "it amazed his doctor and mother when he did that." Although he eventually succumbed to his disease, Turk's quality of life improved drastically. He even obtained his driver's license -- a cherished goal of every teenager. The epilogue to this story is that the cool suit not only gave Turk the strength and ability to live his life more fully, but it helped him to be a happier person.

Moody has also helped a hospital in Norfolk, Virginia, to become equipped with cool suit technology for people who need it. In 1995, a two-year-old girl contracted encephalitis from a mosquito. In need of a cool suit, the Children's Hospital King's Daughter's in Norfolk contacted Moody and asked for her help. She then worked to get the child a suit in only a few days. She later decided that it would be a good idea for the hospital to have this kind of equipment available at a moment's notice. Moody accomplished another milestone in 1995, as the Norfolk hospital accepted an integrated cooling system that can operate 2-4 cool suits at once.

The past accomplishments of Sara Moody and the HED Foundation are truly inspirational. The future also promises to see that even more children are provided with cool suits. The prices of the suits have declined, while the variety of suits has increased. These trends will help more people in the future. The prices for the suits range from $500-$2,000. Six types of suits are now available. They are made by several companies, including ILC Dover, who makes hidden

The light Microclimate cool vest gives the user greater freedom for indoor and outdoor activities. (Courtesy Microclimate Co.)

cool suits for teenagers; Enviromed, the maker of suits for quadriplegics; Durakold, who makes suits that provide partial cooling; and LSSI (now Life Enhancement Technologies), the only suit maker for small children. A newcomer on the cool suit scene is Microclimate of Saginaw, Michigan. The company has already pioneered innovative improvements in cool suit designs by producing suits that are lighter, more compact and easier to use. These improved suits are now in high demand and will add to the quality-of-life for the people who use them.

The last two years, 1995 and 1996, have been marked by change and transition for the HED Foundation and its sponsors. The corporate support of Po Folks Restaurants vanished as the company filed for bankruptcy in 1995. This unfortunate event has left the HED Foundation with dwindling contributions and funds for purchasing cool suits for children who continue to be in need. To fill the gap, Moody has organized and conducted fund-raising to keep some money flowing into her organization. A positive development occurred in April 1996: the Lockheed-Martin Corporation contacted Moody to become involved in the Mission HOME (Harvesting Opportunities for Mother Earth) project. Organized by the National Space Society, Mission HOME is a collaborative effort between the aerospace industry and space activist organizations to show the general public that space activities are worthwhile, necessary and important for humanity's future. Moody participated in the project by traveling around America in 1996. She told people about her experiences with the HED Foundation and how space technology helped the lives of many children. For her efforts, Lockheed-Martin promised to mention Moody and the HED Foundation in all their press releases and to give Moody significant media exposure at all her appearances.

Sara Moody continues her work today and she receives the only payment she cares about: to know that the children with cool suits will have a new chance on life. "It is a wonderful satisfaction", explained Moody. "It is so rewarding to see the faces of the kids and their parents when they first try the cool suit and know it works. Later they respond to questionnaires and send me notes. It's so exciting to read them. I get chills down my spine! It's not just a suit we're giving these kids -- it's new hope. The cooling suits have opened the door, and you have to credit NASA with that, for starting the healing process for the children. Kids with cool suits are just like everyone else."

This chapter will conclude with some of the testimonials that Sara has received from the parents of the children who have benefited from cool suits.[28] The words from these parents are genuine, joyful, and thankful. Remember that these are common people who have suffered through some rather tough times. Their words reflect their gratefulness. Clearly, spin-offs affect not only society at large, but they can also touch the lives of many individuals.

Doris and John Sharret

"Dear Aunt Tootsie, I want you to know what Toby's foot panels mean to me. She has grown up over night. She wants to do it her way, she has a better attitude. She doesn't sit in a chair anymore. It's go all the time. Clothes she has gone wild over, so much she couldn't wear before. She has gone to a Bill Anderson concert, a rock concert, a religious concert. Now she stays for all the ball games. Used to be good if she could stay half a game. She stays for the dance after the game. She has a boyfriend going to her senior prom. Something she would never have gotten to do without her suit. She has been to Dollywood. Had a great time. Been to Nashville, Tennessee. I wish I had this when she was born. Her learning to walk with shoes on was painful. Hot summer days and nights were rough. No more *headaches*. No upset *stomach*. She can eat hot food now instead of it being cold. Thanks for being there when [the] going was rough and tough on her. You saved her *life*. I'll never forget you, you're the dearest."

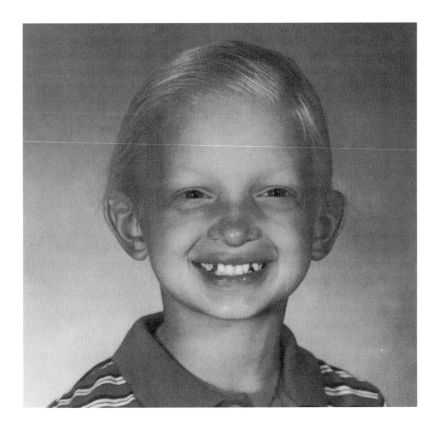

HED cool suit recipient Justin Putrick smiling for his first grade photo in October 1993. (Courtesy Noreen Putric)

Sue Updike

"A simple news clip on television led us into a year-long search for the HED Foundation, Sara Ann Moody and the life-saving cool suit. Even though our family resides in the North Country, we still have summers with sometimes unbearable heat. Josh spent summer days close to home, not participating in any outdoor activities. Since Josh received his cool suit, he is my primary lawn mower, he has attended a Twins baseball game and he rides his bike all over town. The cool suit and the HED Foundation have made the difference for Josh and our family -- the difference between staying home or taking a trip, the difference between activity or confinement -- the difference that means freedom."

Noreen Putrick

"Justin has been ill off and on since the day he was brought home from the hospital. Without the cooling suit, it sometimes might take 4-6 hours to lower his temperature. Justin hasn't had the cooling suit for very long, but we have used it occasionally when we get 80°+ weather. Instead of putting him in a tub bath when his fever rises, we put his cooling suit on and within 15 minutes he's cooler and more comfortable. Justin becomes very ill easily and the cooling suit helps tremendously to bring his fever down.... We're so thankful for the HED Foundation."

Pam Gilbow

"Because of Shannon's skin condition we expect her to be able to participate more normally with the assistance of her new cool suit. She has never been able to play outside in the heat of summer without running extreme temperatures. Her playground time has been limited because of overheating. Now she can have regular recess and playtime with her cool suit. Because of the way the suit cools, she can now disconnect and jump on the trampoline or ride her bike. When she gets hot, [she] reconnects for awhile. We also use Shannon's cool suit to lower her temperature when she runs one. Overall Shannon's new cool suit will become a rich asset to her life as well as to mine. I will no longer have to worry about the heat hurting Shannon."

Chapter 2: Viewing Earth and the Universe

When one thinks of space missions, the most visual and exciting part of is usually the rocket launch. However, after the vehicle is launched and fades from sight observers return to their daily routines. They become preoccupied with the more mundane aspects of everyday life. Most people give little thought to the actual payload of a rocket launch. This is a shame, because once the payload reaches orbit, the mission just begins. Earth-orbiting satellites are undoubtedly the unsung heroes of space activities that have a daily impact on the lives of people around the world. Communications satellites send TV, phone, and fax messages to distant locales. Navigation satellites help people locate where they are and where they are going. Military satellites do the not-so-admirable task of spying on friends and enemies in the never-ending power struggle among the nations of the world. Scientific satellites scan the atmosphere, the land, and the water to gauge humanity's impact on the global ecosphere.

Another type, remote sensing satellites are rapidly becoming a very important tool. Their value is evident as society tries to optimize its use of natural resources and the environment. At the same time, we search for a harmonious balance between the needs of an expanding civilization and those of the natural world. By using the unique Earth-orbital vantage point, remote sensing satellite operators can take a holistic approach when analyzing human impact on the Earth. Conversely, very specific and detailed space observations of a small area permit us to find solutions to a local problem. This space development area is growing rapidly and will guide the use of Earth's finite resources in the future.

The idea of remote sensing simply means the observation or study of a subject from afar. Satellites are in orbit, the Earth is a few hundred miles below the satellites, and hence the term remote sensing satellite came into being. This compares to up-close and direct observation of the object where you may be in direct contact with it or very close by. Take a mountain as an example. You are directly observing and studying the mountain when you are located on it. A remote sensing satellite can study the mountain from space although it is hundreds of miles above, but it has no direct contact with the mountain. This is accomplished by sensors on the satellites that can study features on Earth in different wavelengths of the electromagnetic (EM) spectrum. There is value in viewing Earth in the visible band of the EM spectrum, but a wealth of additional information can be obtained by examining shorter (ultraviolet, etc.) and longer wavelengths (infrared, etc.).

Infrared radiation provides a good example for the value of detecting invisible radiation. It is invisible to the human eye, but can be detected by sensors designed for radiation in the wavelength range of 10-1,000 micrometers. Many objects have infrared radiation because they emit some amount of heat. Humans and all other mammals emit heat as infrared energy. The idea of tracking people in the dark is based on finding that person's infrared signature with specially designed sensors or binoculars. Other types of invisible radiation include short-wavelength X-rays and cosmic rays from exotic celestial objects, and long-wavelength radio waves that we daily listen to for music during the commute to work. Humans can only see a small part of the energy that is being transmitted on Earth and in space. Thanks to historic advancements in science all types of radiation are now visible.

Many vague and general descriptions could be made about the value of remote sensing satellites. It is more instructive to offer specific examples of where and how these satellites are used for an amazing variety of purposes. The first half of this chapter will focus on specific examples of practical work that has been accomplished with remote sensing satellites. Various examples from the trade newspaper *Space News*[29] and from other sources show the versatility and value of these satellites.

The second half of this chapter will highlight pure science. This topic is a departure from the rest of this book, because basic science is not pursued to provide an immediate tangible benefit. The primary reward from basic science is knowledge. This knowledge does, however, usually leads to some practical uses. Assuming no practical purposes are found for some areas of scientific research, how can we justify the accumulation of new knowledge? Can we assign a monetary value on a new discovery? An example is illustrative. What is the value of the Copernican realization that the Earth and other planets orbit the sun? What would life be like, if Nicolas Copernicus had not formulated his ideas and what, if Johannes Kepler had never discovered the relationships of planetary motion? How would we be living today, if confined by an Earth-centered paradigm? In many regards, new scientific knowledge must be assumed to be of great value in today's society. Making any other assumption could have profoundly negative consequences on the future of human civilization.

Space science may also be regarded as somewhat of an unsung hero in the space arena. It is, however, recognized as a very vital activity where new knowledge is being discovered at an ever-increasing rate. The last few years have witnessed an accelerating rate of new findings as orbital and ground-based telescopes have brought a much larger part of the universe into focus. The rate of new discoveries is also likely to continue to increase. Scientists are conducting important and far-reaching astronomical research in space and on Earth. They attempt to answer some of the fundamental questions that were posed since human civilization began: is our solar system unique in the universe? Are there planets around other stars? How old is the universe? Information on research pursuing answers to these -- and other -- questions will be presented to show that knowledge does not have to be practical to have a significant impact on the human condition.

Satellite Imagery Confirms Source of Wetland Pollution[30]

With the aid of satellite imagery, preservationists have determined the source of contamination in a Florida wetlands region. Environmental scientists attributed it to agricultural and urban development.

As a result of their findings, the US federal government filed a lawsuit against Florida in the US Supreme Court in December 1987. The state and the South Florida Water Management District had failed to enforce federal laws regulating the development of land surrounding the Loxahatchee National Wildlife Refuge -- federally protected wetland -- according to environmental researchers. The federal government manages the refuge.

The lawsuit may be settled out of court, however. While the Supreme Court usually handles only appeals, it does handle trials when the federal government is suing a state government. Florida state officials are eager to settle, according to one official.

The problem was discovered through a series of environmental studies, including an inventory of growth in the wetland using SPOT[31] imagery, according to Clark Nelson, director of corporate communications for the US subsidiary of SPOT Image in Reston, Virginia. This subsidiary supplied the imagery.

Researchers at the University of Florida at Gainsville conducted the study, which began in 1985 and continues today. It was requested by the Florida Game and Fresh Water Fish Commission.

Researchers determined that the Loxahatchee refuge was contaminated by runoff water, more than 80 percent of which came from a sugar cane plantation located to its West, according to Florida ecologists. Loxahatchee is near Palm Beach, on Florida's East coast.

Chemical fertilizers and pesticides used in agricultural areas stimulate an unwelcome change in the nutrient value of the wetlands' soil, according to John Richardson, systems economist for the University of Florida at Gainsville.

The refuge borders an urban development on the East. However, ecologists concluded that chemical contamination caused by two water pump stations located on the wetland's eastern edge is a source of only negligible amounts of phosphorous agents that are altering the refuge.

The soil of the northern Everglades, of which the Loxahatchee is a part, is naturally limited in nutrients, according to Richardson. The plants indigenous to the wetlands are adapted to such a growing environment.

However, the high phosphorous content of the agricultural runoff water stimulated growth of vegetation normally not found in the northern Everglades ecosystem. The foreign plant life was cattail and posed the danger of overwhelming an interior wetlands area with a radius of 1-2 km. This constitutes as much as 17 percent of the refuge, or 20,000 of the total 141,000 acres of the Loxahatchee.

Damage affected not only to the plant population in the area, but also the wildlife. The cattail did not promote active wetland wildlife as the indigenous plants, such as sawgrass. Cattail did not attract birds and fish as much as other wetlands vegetation. This pollution of Florida's wetlands was even more severe in other areas, according to ecologists studying the problem.

"The Everglades is a unique ecosystem, and it is being [endangered] by agricultural developers dumping dirty water into the wetlands", according to Randy Kautz, a researcher with the Florida Game and Fresh Water Fish Commission.

Kautz works on a statewide mapping and classification project of Florida's vegetation, using remotely sensed imagery provided by SPOT and Landsat[32] satellites.

Imagery Lends Muscle to Yearly War Against Gypsy Moths[33]

Each spring, the gypsy moths causes considerable damage to forests. This has prompted foresters to use satellite imagery to assess the damage and the effectiveness of various pesticides to control the infestation, according to forestry officials.

The US Forest Service is using satellite imagery of the infested areas from SPOT Image's US subsidiary in Reston, Virginia, according to Clark Nelson, SPOT's corporate communications manager.

The larvae are hairy, inch-long caterpillars that feast on deciduous tree leaves before metamorphosing into moths. Remote-sensing data is effective for tracking the insects, as well as quantifying the damage caused by the gypsy moth larvae, according to researchers.

Using the imagery, researchers classified the forests under investigation as either damaged or undamaged. They followed up with ground surveys to ascertain whether the damage was caused by gypsy moth infestation or by something else, such as logging, according to Bob Acciavatti, researcher for the US Department of Agriculture's Forest Service in Morgantown, WV.

In the satellite imagery, damage from the moth larvae looks the same as that from logging, Acciavatti told *Space News* in a May 9 telephone interview.

Because of this, the satellite images should be supplemented by some aerial photography and ground surveys.

However, the speed with which satellite imagery can be taken benefits the research, because gathering ground and aerial photography is more time-consuming.

The moths are most prevalent on the East coast and affect areas so vast that aerial and field photography are inefficient for cataloging damage. The Forest Service's project to control the insects spans more than 13.5 million acres.

A single satellite scene shows approximately 90,000 acres. Multispectral imagery collected by the SPOT 2 satellite has a resolution of 20 meters (66 feet), meaning objects that size and larger are discernible in the image.

The Shenandoah National Park and the George Washington National Forest, both in Virginia, are two areas for which satellite imagery was gathered to assess the gypsy moth damage. SPOT imagery taken mid-June 1990 showed a considerable amount of defoliation caused by the feeding caterpillars. Imagery of the same sector from mid-July 1990 showed the defoliation had spread. Other imagery was taken of areas being treated by pesticides. One experimental area recorded on SPOT data was a mountain ridge in West Virginia. One side of the ridge was treated with a pesticide, while the other remained untreated. Images show the treated side of the mountain ridge suffered much less damage than the untreated half.

Acciavatti participated in a pilot program between 1987 and 1989 that tested the accuracy of the satellite data.

Although the gypsy moth population is expected to decrease this year, researchers in the Forest Service Atlanta division will continue to use the satellite data to record the effectiveness of certain pesticides. Estimates of the degree of population reduction between 1990 and 1991 may be made, according to government officials.

Satellites Used to Trace Plutonium Runoff[34]

Scientists have developed a satellite imagery process to map drainage patterns in desert climates to track runoff water carrying toxic wastes.

The US Department of Energy (DOE) fully funded the experiment, dubbed Reflex -- the Remote Fluvial Experiment. It was designed to investigate the hydrological cycle in lands managed by DOE, beginning with Nevada's Plutonium Valley.

Satellite remote-sensing imagery was key to the project's success, according to Thomas Gardner, one of the lead investigators.

Plutonium Valley was the site of above-ground testing of nuclear explosives in the mid-1950s. These tests contaminated the site with radioactive plutonium. Plutonium has a half-life of as much as 76 million years, thereby posing a long-term threat to humans. DOE officials were therefore interested in determining the current and future locations of the plutonium fragments.

Satellite imagery proved to be invaluable because of the physical nature of the desert floor. The pattern of water channels cannot be detected by ground surveys, but data from the French SPOT satellite differentiated the rain runoff channels from the surrounding desert floor.

The channels and desert floor differ in mineral composition. This affects the amount of light each reflects and allows satellite sensors to distinguish between them. Because of the desert's arid climate and rocky terrain, rain does not permeate the soil. The multi-spectral imagery helped to identify plant life that might slow the flow of rainwater runoff.

In Plutonium Valley, the runoff channels indicated no past or present danger of the contaminants leaving the test site, according to Gardner. However, the research served as a litmus test proving the usefulness of satellite data. Ground and air surveys provided a plethora of data on the valley. In comparing the data to those gathered by satellite, researchers determined the satellite data could be used with a very high degree of accuracy.

Investigators used both black-and-white 10-meter (33-foot) resolution data and SPOT multi-spectral data with 20-meter (66-foot) resolution, provided by the SPOT Image US subsidiary based in Reston, Virginia. They also used Landsat data with a resolution of 30 meters (99 feet), marketed by the Earth Observation Satellite Company of Lanham, Maryland.

According to Gardner, an associate professor of geology at Pennsylvania State University in University Park, the satellite imagery comprised two of many layers of data in the geographic information system. "This was just the first step in what I hoped would be a 10-year project", Gardner said in a July 31 telephone interview with *Space News*. The study eventually would have included a number of polluted sites around the nation.

Gardner's next goal was to conduct a similar test at 30 Mile Wash, also in Nevada. This too was a test site for nuclear explosives. The contaminants from this site were being washed alongside the Yucca Mountain Range, leaving the DOE's controlled test site.

Plans to conduct this research were thwarted, when additional funding for the study was cut by the White House Office of Management and Budget about two years ago. According to Gardner and the project manager at DOE, Frank Wobber, "the project had the potential to really benefit the DOE. In total, as much as $2 million was invested in the program."

Imagery Aids Wildlife Study[35]

When the California Department of Forestry and Fire Protection wanted to determine the health of wildlife habitats in Northern California, it asked Geographic Resource Solutions to create detailed maps of the area using satellite imagery. Geographic Resource Solutions, founded in 1989 in Arcata, California, specializes in natural resources management.

Although the company had conducted several forest studies using aerial photography, this was the largest forest project and the first one that relied on satellite data, according to Glen Brown, director of remote-sensing applications for Geographic Resource Solutions.

In December 1990, Geographic Resource Solutions won a $450,000 contract from the Forestry Department's Forest and Range Land Resources Assessment Program. The contract directed the company to map six million acres of vegetation in three counties along the Pacific Coast, Mendecino, Humboldt and Del Norte, as well as one million acres in the northern San Joaquin Valley.

This contract is part of an effort by the Forestry Department to develop a wildlife habitat database that will be used to assess timber harvest plans and the long-term health of the ecosystems in the state's forests, according to Janine Stenback, remote-sensing specialist for the Forestry Department in Sacramento.

"This is a pilot program to test what it will take to do it statewide", according to Stenback. The study will observe the different effects of land management on wildlife prosperity. Each map will include information on 20 types of vegetation, the size of plants and trees, their density and whether trees of various sizes are located in the same area.

To create the maps, Geographic Resource Solutions uses satellite imagery from the Landsat 5 thematic mapper, which measures seven types of light in three visual bands to identify different types of vegetation. Landsat remote-sensing satellites are owned and operated by the National Oceanic and Atmospheric Administration. The Earth Observation Satellite Company (Lanham, Maryland) markets the information.*

Geographic Resource Solutions also plans to use aerial photographs of Northern California forest land to check its vegetation classifications.

The company plans to deliver its completed vegetation maps to the Forestry Department by the end of the year. The accuracy of those maps will be evaluated next spring when Geographic Resource Solutions sends people into the forest to gather information on vegetation.

Satellites Used For UN Cambodian Refugee Relocation[36]

A ground-breaking United Nations program using satellite images to determine where to resettle Cambodian refugees was plagued with delays and unexpected glitches. UN officials nevertheless called the effort a success and recommended further use of space technology.

"The use of satellites is a very good planning tool in a place, such as Cambodia, where very little information is available", according to Daniel Mora-Castro, a senior water development specialist, who analyzed satellite data for the United Nations' refugee commission. "For an organization like the [United Nations'] High Commission on Refugees, the use of satellite information is a must."

The satellite imagery speeded up the planned relocation of 350,000 refugees who wanted to leave temporary camps on the Thai border and return to their war-ravaged homeland, according to UN officials. The satellite image maps pinpointed areas with fertile farmland where refugees can resettle.

"If we didn't have access to [the satellite images], we would have had to just use the material compiled by government authorities and spend more time on verification", according to Darioush Bayandor, coordinator of the United Nations' Cambodian repatriation office, in Geneva.

"The satellite enabled us, from laboratories, to identify a much bigger zone and then take a look back at [the area] and check", according to Bayandor, noting it was the United Nations' first use of satellite data from a refugee program.

The French government donated 3.5 million francs (about $630,000) for the SPOT images used by the refugee commission to identify potential areas where Cambodian refugees could resettle.

The UN project encountered delays in fall 1991 when the image maps produced by the SPOT 2 satellite arrived later than anticipated, according to Mora-Castro. The images were expected originally in October and November. They were delivered from early October through early December. SPOT Image and its subcontractor, Geosys, both of Toulouse, France, produced the maps and computer analyses.

To assure that the project's deadline would be met, the High Commission on Refugees arranged to get satellite images of the same region in western and northwestern Cambodia from the Landsat spacecraft, operated by Earth Observation Satellite Company of Lanham, Maryland.

The Landsat data, although less detailed than the SPOT images, enabled UN analysts to begin studying the region while awaiting the SPOT data.

Michael Pousse, applications unit manager for SPOT, explained that technical problems forced SPOT to deliver some maps late, but he maintained that 80 percent arrived on time and that the remaining maps were hand-carried to Cambodia to assure the project's success.

After analyzing both satellites' results, UN technicians concluded that SPOT's images were more useful than Landsat's in determining where to relocate refugees.

UN analysts have to consider more than the availability of fertile agricultural land when determining where to send the refugees, according to Bayandor. Because land mines left from the two-decade war litter much of the Cambodian countryside, the United Nations also sent inspectors to check for the presence of explosives. The mines do not show up on satellite images.

Mora-Castro wrote an internal UN report that faulted some of the Geosys computer-interpreted image maps. Because Geosys based its findings on the only available Cambodian land-use maps, which were decades old, some of the image maps were less accurate than UN officials had hoped.

The Geosys image maps can distinguish between occupied and abandoned tracts of agricultural land, as well as forested areas. The computer analysis marked abandoned farmland in brown and forests in green.

However, when UN technicians checked the image maps against the actual landscape in Cambodia, they found many discrepancies.

Colorado River's Irrigation Value Surveyed Via Satellite[37]

The US Department of Interior's Bureau of Reclamation has turned to remote-sensing data to quantify the value to Southwestern communities of the water flowing from the Colorado River.

By congressional mandate, the Bureau of Reclamation must produce a study once every five years outlining the amount of water in the Colorado River system and the specific benefits it provides for industry and agriculture, according to Bruce Whitesell, a Bureau of Reclamation physical scientist based in Denver.

Since the Colorado River basin is so vast, running through seven states, the study splits it into two independent parts. The Bureau of Reclamation's service center in Denver is compiling information on the upper basin of the Colorado River that includes portions of Colorado, Utah, Wyoming, New Mexico and Arizona.

Meanwhile, the Bureau of Reclamation office in Boulder City, Nevada, awarded a contract to Pacific Meridian of Emeryville, California, in May to map the lower Colorado River basin, which includes land in Colorado, Nevada, Arizona and California.

Under the two-year contract, Pacific Meridian also will teach the Bureau of Reclamation officials the techniques needed to perform their study using remote-sensing data in future years, according to Pacific Meridian President Cass Green.

In the past, the Bureau of Reclamation attempted to compile data on crops irrigated by the Colorado River using agriculture statistics for surrounding counties. However, it was difficult to determine how much agriculture in each county was irrigated by water from the river.

To make the study more accurate and less time-consuming, the Bureau of Reclamation is attempting to develop a computer database showing all the irrigated land. That database will rely on aerial photography and verification by ground surveys to pinpoint the location of all the irrigated land.

Satellite imagery will be used to detect changes over time in the amount of land being irrigated, as well as to derive information on the crops being grown.

The Bureau of Reclamation service center in Denver uses data gathered by the Landsat thematic mapper to determine what types of agricultural products are being grown in various fields in the upper river basin. Eosat Company of Lanham, Maryland, operates Landsat.

"To a certain extent, you can use remote sensing to determine the crop and how wet a year it was", noted Diane Williams, a Bureau of Reclamation geological engineer. "We are just starting to use satellite imagery for 1993 field verification", she added.

In its study of the crops in the lower Colorado River basin, Pacific Meridian will have access to imagery from the Landsat thematic mapper, SPOT Image's multi-spectral and panchromatic sensors, as well as to data derived from an airborne hyper-spectral scanner developed by TRW Space & Electronics Group of Redondo Beach, California. SPOT Image is based in Toulouse, France. "We are evaluating all the imagery to see what works", said Green.

Landsat's thematic mapper data has a spatial resolution of 30 meters (99 feet) in which land features that size or larger are visible. SPOT's black and white imagery has a 10 meter resolution (30 feet). However, Landsat has a greater spectral resolution, and gathers data in a greater number of bands in the electromagnetic spectrum.

Pacific Meridian is gaining access to many different types of imagery because the task at hand is difficult. The Bureau of Reclamation wants detailed information on the crops being grown on every acre. For example, it wants to know where wheat is being grown as opposed to rye, two crops that are difficult to tell apart using remote-sensing data. Moreover, the irrigated land in the lower Colorado River basin is so rich, that much of it supports four crops every year.

The TRW hyper-spectral scanner is a tool that may aid the Bureau of Reclamation in future studies of the Colorado River basin. In June, TRW won a $59 million contract from NASA to include that scanner, which will photograph the Earth in 384 visible to near-infrared spectral bands, on the 700-pound Lewis spacecraft scheduled for launch in 1996.

By capturing images in hundreds of spectral bands, the new scanner may make it easier to use remote sensing to tell the difference among similar crops. The information derived from the scanner may be so detailed, that it may allow companies, such as Pacific Meridian to detect the difference in light reflected by young corn as opposed to older corn.

Scientists Use Landsat to Hunt Ticks in Northeast US[38]

Scientists at the NASA Ames Research Center in Mountain View, California, are using satellite imagery combined with computerized maps to detect populated areas with a high risk of Lyme disease transmission.

To date, the research has focused on suburban areas of New York's Westchester County. In the future, however, the Ames scientists hope to develop techniques that will help them evaluate the risk of Lyme disease transmission in other communities in the northeastern United States.

Infected deer ticks transmit to people Lyme disease, which can lead to arthritis as well as neurological and cardiac disorders. Eighty percent of US cases are reported in the Northeast, according to Dr. Durland Fish, director of medical entomology at the New York Medical College in Valhalla, NY.

Ames scientists are working with officials from the New York Medical College and the Westchester County Health Department, to determine whether satellite imagery is useful in identifying communities or individual properties at greatest risk from the disease.

"We are using satellite imagery to find the habitats where ticks live and the environments where people come in contact with ticks", noted Byron Wood, a remote-sensing specialist working at Ames under a contract with his employer, Johnson Controls World Service Inc.

An initial field study, begun in 1992, relied on data of the Lyme disease antibodies in blood samples from dogs in Westchester County. When Ames scientists compared data from the Landsat thematic mapper and computerized maps with the results of the dog study, they found a significant correlation between the dogs' exposure rate and the proportion of vegetated residential areas located next to woods.

Now the Ames team is taking a closer look at two Westchester county communities, Armonk and Chappaqua, to determine whether they can predict the prevalence of ticks carrying Lyme disease near individual properties, related Sheri Dister, a geographic information systems specialist. Dister is also a Johnson Controls World Service employee working under contract at Ames.

Field workers are dragging a corduroy cloth over the ground and vegetation to determine the density of ticks on individual pieces of property. Scientists, meanwhile, are examining Landsat imagery to determine whether the tick density can be correlated to the amount of vegetation or moisture. The latter can be detected using remote-sensing data from the Landsat spacecraft.

Dister hopes to complete that study by the end of the summer. At that point, scientists in New York will inspect various properties to measure the prevalence of disease-carrying ticks, testing the accuracy of the study's findings.

In the past, medical researchers determined the risk to local populations of contracting Lyme disease by sending entomologists to an area and sampling for ticks. That approach was inefficient because it was difficult to cover large areas with such a labor-intensive approach, and the sampling could only occur during the summer months when ticks were most prevalent.

"Remote sensing is a more efficient method for doing research", according to Fish who believes that the techniques could help the medical community determine the risk of other insect-borne diseases, such as malaria and yellow fever. "Remote-sensing technology provides you with insight into the local environment and these insects are very much adapted to those local features."

Chinese Use Japanese Satellite To Predict Earthquakes[39]

Last June, remote-sensing researchers here [Beijing] began to notice unusual temperature increases off the southwest coast of Hokkaido, Japan's northern island. Two weeks later they predicted an earthquake stronger than 6 on the Richter scale would occur in the region within a week, using data from the Japanese Earth Remote Sensing Satellite (JERS-1).

On July 12, a powerful quake measuring 7.8 on the Richter scale rocked Hokkaido. The on-the-mark prediction was the product of an attempt by Chinese scientists to use satellites to forecast earthquakes. The forecasts include the approximate time, strength and location of a quake.

"We don't have a very good theory to account for this -- we are still at a very early stage", according to Liu Cheng, deputy director of the operations and services division of the Satellite Meteorological Center. As a result, the Chinese team did not publicize its Hokkaido prediction. "According to international practice, you can't forecast earthquakes in other countries because it could lead to social chaos", he added.

Liu's team has made 20 predictions, accurately forecasting 12 earthquakes. The eight remaining forecasts include two that were considered unsuccessful, even though earthquakes occurred, because the quakes happened outside the time limits set by the team for a successful forecast.

Liu's tentative theory holds that tiny fissures appear on the ocean floor before a quake. Gas rises through the fissures to the surface where it meets the Earth's magnetic field. The heated gas hovers in the lower atmosphere, but is difficult to detect without the help of satellite data.

Instruments aboard JERS-1 can measure minute changes in temperature just above the ocean's surface, and these measurements provide clues to impending earthquakes, Liu believes. The warmer the temperature, the stronger the impending quake, he theorizes.

JERS-1 was launched in February 1992 aboard an H-1 rocket. Its principle instrument is a synthetic aperture radar for all-weather surface imaging, but the satellite also carries a suite of optical sensors that detect infrared wavelengths.

Liu's research has intrigued some Earth scientists in the United States. "The work is pretty unique", according to Louis Walter, associate director for Earth sciences at Goddard Space Flight Center in Greenbelt, Maryland. "But there are a lot of people skeptical about it -- the causal chain is a little weak."

The gaseous emission and Liu's theory about its interaction with the Earth's magnetic field are too little understood, Liu noted in a February 21 interview.

Walter, who also chairs the committee on space and natural disaster reduction for the International Astronautical Federation, proposed in November that Liu's team make a forecast monitored by the organization. The forecasts would not be publicized, but would be strictly held for scientific evaluation, Walter promised. Liu has not responded.

The team in Beijing also predicted what they thought would be an earthquake in the northern Philippines, just before the eruption of Mount Pinatubo in 1991, a research team member indicated. That could be evidence that the theory might prove effective in predicting volcanic eruptions as well as earthquakes, she added.

Operating on a shoestring budget equivalent to a few thousand dollars a year, Liu's team continues to gather data on the theory. "Of course, we are still in a research phase, and we expect failures as well as successes", one member of the team observed.

"I wish I had had data before the Los Angeles earthquake", Liu added, referring to the quake that rocked that city in January. "I'm very worried about the situation there."

Imagery Helping To Determine Rain Forest Depletion Rate[40]

A new NASA study of satellite imagery promises to nail down an elusive and politically potent figure -- the rate at which the world's tropical rain forests are being destroyed.

Researchers at the University of New Hampshire and the University of Maryland may end up revising environmental history by conducting, what they call the first wall-to-wall study of deforestation in the tropics of South America, Asia and Africa. Previous studies extrapolated deforestation rates from a few samples ; this study examines Landsat and other satellite data throughout each region.

Bill Lawrence, director of the project at the University of Maryland, pointed out that in the past, researchers mistakenly relied on ground survey samples that did not accurately reflect the rate of deforestation. "This project has shown the only way to get a really good [deforestation] rate is to observe it directly with satellite data."

So far, the study suggests that forests in Brazil, for example, are being destroyed at a rate of 15,000 km^2 (6,000 square miles) per year. That compares to previous estimates of 45,000 km^2 (18,000 square miles) per year.

Researchers temper this good news with findings that extensive human use is emptying the forests of their trademark treasure -- rich diversity of species. "The greenhouse situation is in better shape because the [deforestation] rate is lower" than suggested by previous research, according to Compton Tucker, a visiting scholar at the University of Washington in Seattle. "But in terms of biodiversity, there is no cause for optimism. There is much cause for concern."

The US researchers check their data against information from forest experts in each area. Through the combination of satellite and ground data, the researchers are creating a profile of the tropics that shows patchy deforestation and rejuvenation of forests stripped decades ago for farmland.

"In many cases the traditional tropical land use is one of short-term cultivation and abandonment and reuse after several decades", Lawrence observed. "In most cases as soon as it's abandoned, unless the soil has been completely destroyed, the forest does come back."

"The forests that grow back, however, are not as rich in biodiversity as the virgin forests", according to Tucker. At the same time, road-building, hunting and other human activities are breaking up the forests into isolated patches.

Through this pattern, biodiversity is shrinking at a rate commensurate with a much higher rate of deforestation. "Generally, the smaller the area, the smaller the biological diversity", Compton noted. "You tend to break off parts of populations, and often they become extinct."

The NASA study is designed to serve as a prototype of the data application and distribution envisioned for NASA's Earth Observing System, a constellation of satellites set for launch starting in 1998. Eventually, NASA hopes to make the information widely available over a computer network to noncommercial researchers, who could buy the data for a modest fee.

The study expands research in Brazil to rain forests throughout the world. The researchers are relying primarily on data from Landsat 4 and 5 satellites. In Africa, where there is no ground station to receive Landsat data, the researchers hope to rely on supporting data from the Advanced High-Resolution Radiometer aboard National Oceanic and Atmospheric Administration polar-orbiting weather satellites, and the French-owned SPOT satellite.

Tucker believes that data from SPOT are more expensive and less valuable scientifically than Landsat images. "Everyone is deeply concerned about the continuation of the Landsat program. Without this, we will lose a tremendous information source."

David Skole, who is leading the project at the University of New Hampshire, believes the study will be one of the first comprehensive applications of Landsat imagery by researchers. Researchers need a continuous supply of Landsat data. "It would be an embarrassment that we have come this far -- to using Landsat for this kind of work -- only to lose the program."

Shuttle Radar Sensor Used to Study Gorillas in the Mist[41]

Wrecked by ethnic war and almost permanently shrouded in clouds and mist, the habitat of Central Africa's endangered mountain gorilla is impervious to traditional satellite and airborne mapping and observation methods.

Scientists mapping the gorillas' homeland can now perform their work with the aid of imagery gathered by a space shuttle-borne radar sensor. The imagery from the $320 million Spaceborne Imaging Radar C/X-band Synthetic Aperture Radar device was sent by computer from NASA's Jet Propulsion Laboratory in Pasadena, California, to Rutgers University's Center for Remote Sensing and Spatial Analysis.

Field researchers who used Global Positioning System (GPS) receivers will combine the digital radar image with other data logged. The receivers use the 24-satellite GPS navigation constellation to gather location information to create a three-dimensional, computer-based geographic information system that charts the gorillas' homeland.

The project is a joint effort between Rutgers University, New Brunswick, NJ, and the Dian Fossey Gorilla Fund. Rutgers anthropology professor H. Dieter Steklis is executive director of the gorilla fund that was created in 1985. The fund carries on the pioneering research work of Dian Fossey, whose efforts to study and protect the animals were chronicled in the movie "Gorillas in the Mist."

"Roughly half of the estimated 600 to 650 mountain gorillas in the wild today live in the Virunga Volcano chain that straddles the borders of Rwanda, Zaire and Uganda", according to Steklis.

Once the basic database is completed later this year, researchers will better understand the gorillas' movements and why they only live in certain areas of the volcano chain, according to Greg Movesian, a representative for the gorilla fund. The database also will help the fund keep track of poaching activities and the impact of human encroachment.

Clouds or rainy weather frequently cover the region, making observation with traditional optical sensors virtually impossible. Since radar can penetrate the cloud cover, the radar images will be "of immense value to the project", according to Scott Madry, associate director for the Center for Remote Sensing and Spatial Analysis at Rutgers University.

"Radar isn't affected by the weather; it can penetrate the mist", noted Madry, a former NASA employee who worked at the agency's Stennis Space Center in Bay St. Louis, Mississippi. "I begged and pleaded with NASA that we have a desperate need only they could help us with", he added that NASA and the Jet Propulsion Laboratory "have bent over backwards to help us protect these magnificent creatures."

Diane Evans, scientist for the C-band radar project at the Jet Propulsion Laboratory, noted that NASA officials approved Madry's request because of the project's unique circumstances and worthy goals. However, the Central African images were taken "on a non-interference basis with the primary mission", according to Evans. The data were processed back at the laboratory at no additional cost to the program as part of the radar team's engineering and equipment check-out phase.

For Madry the space-borne radar data were "our only hope of getting comprehensive coverage." The ethnic war between the Hutu and Tutsi tribes in Rwanda would have made it unsafe for aircraft to fly over the region. He added that "getting government approval to fly over all three countries at once would be impossible."

Parts of the region to be included in the study, especially in Zaire, have never been accurately mapped before, while other areas have not been mapped since the 1950s. The radar image will update existing maps and give researchers their first detailed look at the region's terrain and vegetation.

In all, the radar device took about 14 images on two orbit passes over Central Africa. Although he has received just one of those images, Madry hopes to obtain the rest later this year. He also requested similar remote-sensing images during the device's second flight about the shuttle Endeavour in August. NASA officials will try to accommodate Madry again.

The team hopes to return to Rwanda once the ethnic war is over so they can continue obtaining GPS references on the gorillas' movements.

Space Radar Studies Archeological Site In Cambodia[42]

Images from the international Space Radar Laboratory (SRL) may help researchers find previously unknown settlements near the ancient city of Angkor in Cambodia.

The radar data were obtained during the October [1994] flight of NASA's Space Shuttle Endeavour, processed and sent to the World Monuments Fund (WMF) in January [1995]. The group had approached the radar science team about observing the Angkor area after SRL's first flight in April 1994.

"I had read about the radar mission while the April flight was in progress and instantly surmised that it would have applications to the international research efforts at Angkor", according to John Stubbs, program director for the fund. "I didn't really know where to start, but I was hopeful NASA would be willing to image the area around Angkor."

Angkor, a vast complex of more than 60 temples dating back to the ninth century AD, served as the spiritual center for the Khmer people. At its height, the city housed an estimated population of one million people and was supported by a massive system of reservoirs and canals.

The April flight of SRL's complementary radars, the Spaceborne Imaging Radar-C/X-band Synthetic Aperture Radar (X-SAR), first demonstrated their capability to obtain vast amounts of data applicable to ecological, oceanographic, geologic and agricultural studies.

"We realized after the huge success of the first flight that we could be more flexible in adding new sites to the timeline of flight two", noted Dr. Diane Evans, the SIR-C project scientist at NASA's Jet Propulsion Laboratory, Pasadena, California. "Since our science team was interested in studying as much of the tropical rain forest as possible, Cambodia and the Angkor site seemed to be a great complement to our ecology objectives."

Today, a dense rain forest canopy hides Angkor. Weather, war and looters have ravaged its temples. Its extensive irrigation system has fallen into disuse.

"The radar's ability to penetrate clouds and vegetation makes it an ideal tool for studying Angkor", observed Stubbs. "I can see the canal-and-reservoir system very clearly in the radar imagery, and preliminary analysis reveals what may be evidence of organized settlements of large tracts of land to the north of the present archeological park, which until now, has gone unnoticed."

The SIR-C/X-SAR data will be used by the WMF, the Royal Angkor Foundation and research teams from more than 11 countries to understand how the city grew and then fell into disuse over 800 years.

"The 'temple mountain' monuments of Angkor, such as Angkor Wat and the Bayon, are not unlike some of the pyramidal forms encountered in Central America", Stubbs said. "The sheer size and sophistication of Angkor's great city plan, now enveloped in dense jungle, sets this ancient capital apart as the ultimate jungle ruin."

SIR-C/X-SAR is a joint mission of the United States, German and Italian space agencies. JPL builds and manages the SIR-C portion of the mission for NASA's Office of Mission to Planet Earth.

The Compton Gamma Ray Observatory

NASA's Arthur Holly Compton Gamma Ray Observatory (GRO) is the second spacecraft to make it to orbit in the space agency's Great Observatories program. Separated from the shuttle *Atlantis* on April 5, 1991, the GRO is dedicated to searching the skies for gamma ray emissions. This will eventually lead to the creation of a sky-map of all the observed gamma ray sources. Gamma rays are high-energy radiation that have very low wavelengths and are extremely lethal to living creatures.

In its first five years of operation, GRO has made several interesting discoveries in the field of gamma ray astronomy. Gamma rays have been observed from all parts of the celestial sky. Earlier astronomers had speculated that gamma rays only originated from within the Milky Way Galaxy, but apparently gamma rays have been detected beyond our galaxy as well.

The GRO has also discovered a new source of gamma radiation, called gamma ray quasars. Quasars are thought to be very distant and faint galaxies formed near the beginning of the creation of the universe. The gamma radiation in these quasars is thought to originate from the bright cores of these quasar galaxies. Another gamma ray discovery has been made in a different type of galaxy -- Seyfert galaxies. Seyfert galaxies are spiral galaxies, such as the Milky Way, but have an unusually bright nucleus or central region. It has been discovered that these galaxies emit gamma radiation at lower energy levels than previously thought. This provides some evidence that Seyfert galaxies may be a source of diffuse gamma rays.

One of the most interesting discoveries of the spacecraft involved the observation of a celestial object that acted like no other seen before. The GRO found an object in December 1995 that exhibited very intense bursting emissions of gamma rays. Within the first day of observations, 140 gamma ray bursts were recorded. The rate of these bursts later settled down to 20 per day. The perplexing discovery was that the source also emitted gamma radiation continuously at a rate of two pulses per day. No other previous source in the sky had been found to emit gamma rays in bursts and in a continuous fashion.

Astronomers were at first puzzled what this object might be, but both types of emissions were found to be coming from the same object. The object in question is a binary star system that includes a pulsar and a low-mass companion. Observations show that the pulsar orbits its companion every 12 days. Continuous emission at a nearly constant rate, in this case two pulses per day, is a classic sign for the discovery of a pulsar. A pulsar is a rotating neutron star that gives off a consistent pulse of energy due to its rotation. A neutron star is the incredibly dense and collapsed remnant of a moderately large star that has exhausted its supply of hydrogen and helium fuel.

Also involved in the initial observations of this star system was NASA's Rossi X-ray Timing Explorer, launched on December 30, 1995. This satellite has been important in viewing these celestial object because they emit both X-rays and gamma rays. Rossi satellite observations confirmed that they sent bursting and continuous and that this source was bright in a range that varied from 2000-60,000 electron volts. Astronomers suggest that the pulsar's emissions are caused by the impact of gaseous material from the low-mass companion that is drawn to the pulsar by its overwhelming gravitational attraction. The matter violently collides with the pulsar at speeds up to half the speed of light. This impact consequently heats the pulsar's surface to temperatures over 1,000,000,000°F.

The Hubble Space Telescope

Probably the most well known and, for a time the most maligned, NASA project, the Hubble Space Telescope (HST) is now proving its value to the astronomical community, following a repair mission by the crew of the shuttle *Endeavour* on STS-61 in December 1993. Originally launched on April 24, 1990, HST was soon found to exhibit a well-known optics flaw called spherical aberration. The slightly incorrect shape of Hubble's primary mirror did not allow the incoming light to be focused at a specific point. This caused a halo effect and hampered the spacecraft significantly. However, with the repairs successfully completed, astronomers are taking advantage of Hubble's abilities and will continue to do so throughout the 15-year lifetime of the science satellite.

At Hubble's five-year anniversary, project and program scientists intimately involved with the telescope, compiled a list of its top 10 accomplishments based on their scientific merit and their long-term importance in advancing the field of astronomy. The list includes:

• Offered the first conclusive evidence for the existence of immense black holes, millions or billions of times the mass of Earth's Sun.

• Showed that the universe might be younger than had been previously thought.

• Gave the first direct visual evidence that the universe is evolving as predicted in Big Bang cosmology, by resolving the shapes of the farthest galaxies ever seen.

• Discovered that quasars, very distant and remarkably bright objects, are even more mysterious than commonly thought because many do not dwell in the cores of galaxies, but are isolated in space.

• Suggested that dark matter in the universe is more exotic than previously thought, by finding that Nature does not make enough of the extremely small Red Dwarf stars that were once a leading candidate for the universe's "missing mass."

• Supported the Big Bang theory by refining estimates of the amount of deuterium in space, an element created in the initial cosmic fireball that gave birth to the universe.

• Solved the mystery of intergalactic clouds of hydrogen by showing that they are very gigantic halos of galaxies.

• Implied that planets, and presumably life, might be abundant in the universe by discovering disks of dust that might be embryonic planetary systems around young stars.

• Provided important details and surprising findings of the spectacular collisions of comet Shoemaker-Levy 9 with Jupiter in 1994.

• Revealed dynamic weather changes on nearly all the planets with such clarity once attainable only with spacecraft flybys.

From this list, it is fair to say that the general public is often interested in the continuing debate about the age of the universe and the possibility that planets (and life) exist around stars other than our sun. In both instances, HST has made significant contributions in getting answers to these age-old questions.

One of the original goals of the HST program was to use the telescope to make a more accurate age estimate of the universe. According to Big Bang cosmology, the universe began as an incredibly small, dense concentration of matter that cataclysmically exploded outward. Today, astronomers see the expansion of the universe occurring as galaxies in all parts of the sky are receding away. The velocity at which objects are moving away from our point of observation is measured by means of Doppler shifts. If an object is moving away from us, it exhibits a wavelength shift into a longer-wavelength region. The goal of the Hubble Space Telescope Key Project team is to find a more accurate value for the Hubble constant. Without discussing the physics or math involved, the Hubble constant is a measure of the rate of expansion of the universe. The value for the Hubble constant is also inversely proportional to the age of the universe. As the Hubble constant goes up, the age of the universe declines.

Reporting on their initial results in 1994 and again reporting findings in 1996, the Key Project team is coming up with values of the Hubble constant that are fairly similar. Two teams of astronomers involved with this project are using different methods to come up with values for the Hubble constant. The first team, led by Wendy Freedman of the Carnegie Observatories in Pasadena, California, reported in 1994 a Hubble constant value of 80 (± 17) kilometers/second/megaparsec.[43] This value derived from the measurement of 20 Cepheid variable stars in galaxy M100 in the Virgo Cluster that is approximately 56 million light-years away and red-shift measurements from the Coma

Cluster. Her team reported in May 1996 that five ways of measuring the Hubble constant yielded values from 68-78 kilometers/second/megaparsec.

The second team, led by Allan Sandage of Carnegie Observatories, uses a single method to measure a value for the Hubble constant. Their observations are based on the intrinsic brightness of Type IA supernovae. Sandage suggests that all Type IA supernovae will reach the same brightness level and are thus an excellent celestial object to use to measure distances. Because supernovae are very bright, it is thought that they will be visible further back in time than by other distance indicators, such as Cepheid variable stars. The Sandage team reported a value of 57 kilometers/second/megaparsec for the Hubble constant in March 1996.

The Hubble constant values correspond to universe ages of 9-12 billion years and 11-14 billion years, respectively. The controversy right now involves the seemingly contradictory observation that the estimated age of some stars in our galaxy is greater than the age of the universe in some of the estimates for the Hubble constant. The Key Project aims at measuring the Hubble constant to within 10 percent of its true value.

Protoplanetary disks and other solar systems

Space- and ground-based telescopes have provided increased evidence over the last 10 years for the existence of planets around stars other than our own. The first piece of evidence involves the discovery of disks of dust located around young stars that are thought to lead to the formation of planets and whole solar systems. From the work of both HST and ground-based telescopes, it is apparent that disks of dust are common around young stars in the universe. Two examples of this include an examination of stars around the Orion Nebula and a detailed look at the star Beta Pictoris.

The Orion Nebula is an area of star formation located within a much larger molecular cloud that is 1500 light-years from Earth. If current theories of solar system formation are correct, then dust disks should be expected around many young stars in this nebula. Both before and after its optics were fixed, the Hubble telescope focused on the Orion Nebula. With its uncorrected optics flaw, the telescope observations in 1991 could not verify that dust disks were present around the stars, but there was enough clarity to suggest that possibility. Hubble again focused on the nebula in 1994 after its optics were corrected. The much clearer second image displayed fascinating results and beautiful pictures of a nursery of young stars many of which were surrounded by the expected dust disks. In the area of the Orion Nebula observed, 56 of the 110 stars observed were found to have large quantities of dust surrounding the stars. These, along with other observations, are significant evidence that there may be innumerable solar systems in the Milky Way and throughout the universe.

Specific observations have been made of the star Beta Pictoris. At least one study provided indirect evidence that planets may be orbiting this star.[44] Conducting their research at the European Southern Observatory (ESO) in La Silla, Chile, astronomers P.O. Lagage and F. Pantin used a special camera to observe Beta Pictoris from January 6-11, 1993. Their results not only showed the presence of a protoplanetary disk of dust, but also certain irregularities that hint at the possible presence of planets.

The overall disk was discovered by observing the 10 micrometer region of the electromagnetic spectrum. Protoplanetary disks are identified by their excess energy emission in the infrared that is caused by hot dust grains within the disk. When observing the star in the infrared, the astronomers noted the brightness of the region around the star and determined it to be a dust disk.

The two most important characteristics of this dust disk were its asymmetrical shape and the decrease in the amount of dust present within 40 astronomical units (AU) of the star.[45] The asymmetrical shape may be a clue that planets are present because even a slightly eccentric (non-circular) orbit would alter the shape of the dust disk. The lower density of dust in the inner portion of the disk is an important finding. Astronomers believe that newly born planets would be very effective in sweeping out a path through the disk along their orbital path. These planets could also possibly clear the region inside their orbit of most of the dust in as little as one million years. With this and evidence that includes the detection of silicate particles within the disk, the authors suggest that a planet may have formed at a maximum orbital distance of 20 AU from Beta Pictoris.

The discovery of extrasolar planets

Besides the detection of protoplanetary disks around young stars, we have witnessed in the last 10 years what is currently thought to be the first planets outside our solar system. The discovery of other planets is intimately associated with humanity's interest in the possibility of life existing elsewhere in the universe. If protoplanetary disks are found, then planets would be expected to form. If other planets are discovered, then the question arises about their ability to sustain life. Although extrasolar life has not yet been detected, the discovery of several different possible planetary systems since 1990 increases the chance that life may exist elsewhere in the universe.

Although there were several announcements of extrasolar planets that were later retracted (Barnard's Star, 61 Cygni, Lalande 21185), there is significant evidence to suggest that three Earth-mass planets are orbiting the millisecond pulsar PSR B1257+12[46] and that a single Jupiter-like planet is orbiting the star 51 Pegasi.[47]

Detailed efforts were undertaken to ascertain that the observations of PSR B1257+12, detected first in 1992, were planets and not some other phenomena. The method used to detect the planets involved studying the rotation rate of the pulsar. Pulsars are neutron stars that spin at a nearly constant rate. If planets orbit this pulsar, they would exert a gravitational influence on the pulsar that will slightly affect the pulsar's rotation rate. Alexander Wolszczan from Pennsylvania State University confirmed in 1994 that the pulsar perturbations derived not from any extraneous source, but likely from the presence of planets. All three planets are less than 1 AU from the pulsar and have approximate masses of 0.015, 2.8 and 3.4 times the mass of the Earth. It is highly unlikely that any life could exist on these planets that now orbit a dying star.

Michel Major and Didier Queloz of the Geneva Observatory have presented quite a different scenario for the presence of a planet or a stripped brown dwarf around a star that is quite similar to our sun.

51 Pegasi b is a star similar to our sun, but is about 10 billion years old with a composition that includes an overabundance of heavier elements. The object discovered, called 51 Pegasi b, has an estimated mass of 0.5-2 times that of Jupiter. It orbits very close to the star -- at a radius of only 8 million kilometers. This proximity generates a surface temperature for 51 Pegasi b of more than 1300° Kelvin. This object was detected by measuring changes in the radial velocity of 51 Pegasi b. These velocity changes resulted because the two objects orbit around a common gravitational center. 51 Pegasi b can be seen to alternately move toward and away from our line of sight on Earth.

The researchers suggest that the object is a stripped brown dwarf. A brown dwarf is an object with insufficient mass in its formation to begin the nuclear fusion that leads to the creation of a star. Brown dwarfs and planets, such as Jupiter, differ also in size. Jupiter has a composition of a star, with mostly hydrogen and helium, but is too small to classify as a brown dwarf. Either way, the gravity of 51 Pegasi may have drawn much of the mass away from 51 Pegasi b to bring its current mass down to that near Jupiter. There is also the possibility that another planet exists in this system and research is underway to locate it.

Other possible planet discoveries have also been recently announced. Astronomers Geoffrey Marcy and R. Paul Butler have discovered two planets that orbit around different stars and resemble Earth's sun.[48] Both 70 Virginis in the constellation Virgo and 47 Ursae Majoris in Ursa Major (the Big Dipper) have Jupiter-like planets orbiting them. The planets were detected in a seven-year search of 120 stars, using a spectrograph mounted to a three-meter telescope at California's Lick Observatory. Discovered by its periodic radial velocity changes, the planet around 70 Virginis is estimated to be eight times more massive than Jupiter. The planet around 47 Ursae Majoris is about 2 AU from its sun and has an estimated 3.5 times greater mass than Jupiter.

Chapter 3: Costs and Benefits of Spaceflight

United States citizens often engage in hypocrisy when it comes to discussions about the federal government and federal spending. Built upon the ideals of democracy, political representation, freedom of speech and association, the citizens of this country have essentially been given carte blanche to live their lives as they see fit since 1776. Remarkably, however, many Americans -- past and present -- see their federal government as an impediment to their personal and professional success. This view is common everywhere today as many business people complain about government regulations. US presidential candidates perpetually call themselves "outsiders" of the "establishment" and claim to work for the average person. Ordinary citizens constantly complain about having to pay taxes. On the other hand, people loudly protest when government programs are cut that directly affect them. Many individuals accept the basic premise that free enterprise is good and big government is bad. However, this skewed assumption ignores the negative aspects of business and the many benefits of government.

As with any discussion about government, many people do not know how their government works or how their money is being spent. In this chapter, the NASA budget will be placed into proper perspective with the entire federal budget. When seen in this context, NASA has received very little funding. This is one reason why we are not on Mars yet. NASA today receives less than one cent for every dollar the federal government spends. NASA's effect on the US economy will also be examined, using a few studies that show the economic effects of procurement spending and spin-offs.

There is another segment of the space program that most have never heard of before. The "other face" of the space program is the commercial space industry. Most people who do not closely follow space activities, would incorrectly assume that NASA builds, launches, and operates all the spacecraft in the United States. This notion is, however, incorrect. NASA is rarely responsible for building the hardware it uses to send people and payloads into space. The space shuttle is a good example of hardware not made by NASA. Rockwell International in Palmdale, California, has been the sole builder of orbiters for NASA under contract. Although NASA does build a few spacecraft with agency personnel, typically robotic probes, the standard situation is for

NASA to choose a contractor to build a particular vehicle. There are also other private space efforts that have virtually no contact with NASA at all. These entrepreneurial efforts work to use venture capital and other private money to kick-start their space ambitions. The second half of this chapter provides a survey of commercial space activities and a sampling of the different efforts that will become a part of many people's lives in the future.

The NASA Budget

Since the end of the Apollo era, it can be accurately said that space exploration and development have not been a priority of the United States. NASA limped through the 1970s with only the Skylab space station, the Viking probes to Mars, and difficult space shuttle development. Project Apollo was unwisely eliminated altogether after the 1975 Apollo-Soyuz linkup in orbit. This decision kept Americans out of space for almost six years. The 1980s were dominated by a space shuttle program that had its share of successes and failures. On the negative side, the shuttle system operation proved less economical than originally promised, and we suffered through the terrible tragedy of the *Challenger* accident in 1986. On the positive side, NASA successfully rebounded from *Challenger* and now operates a launch system that conducts valuable scientific research, rescues and repairs orbital spacecraft, such as the Hubble Space Telescope. It has opened up access to space to more people than any other launch system. However, the shuttle is operated as an end in itself with its only other significant mission being the construction of the international space station beginning in 1997 or 1998. The single most significant problem with the US space program, is that it does not have a direction or goal in which to focus its space-based activities.

We can firmly blame NASA's lack of direction on the past and present occupants of the White House and the Congress. It is the responsibility of each US president to enunciate a space policy that will guide NASA in its activities. It is also the responsibility of the Congress to fund the space program so that it can accomplish its goals. Unfortunately, since John F. Kennedy no president has given the space program a direction that the Congress has agreed to fund.

The best example of a failed effort at providing direction was President George Bush's proposed Space Exploration Initiative in July 1989. This would have permanently returned people to the Moon and initiated expeditions to Mars. The initiative died quickly when Congress refused to fund the effort. The proposal did not seem to find public support. Most people thought the NASA plan was too expensive. If average citizens and elected politicians are funding space activities they need to have good reasons for the necessity of space development. One also needs to understand the cost of major programs compared to the overall federal budget. The next chapter will give some compelling reasons why America and the other nations should be venturing into space.

President Bill Clinton proposed a NASA budget of $14.26 billion for fiscal year 1996.[49] Assuming that these numbers are correct, and that Congress will pass the budget, NASA's share of the total fiscal year 1996 budget will be less than one percent of all US government spending. In other words, NASA receives about one cent for every dollar spent by Uncle Sam. A historical look at NASA's budget also reveals that even during the Apollo program the government spent less than 5 percent of its budget on the space program in its peak funding year. Figures 1 and 2 display past NASA budgets in current dollars and NASA's budget as a percentage of the total US budget.

As evidenced in Figure 1, NASA's budget quickly increased early in the agency's history to meet the demands of Projects Mercury, Gemini, and Apollo. NASA's budget during its first full decade of existence peaked in 1966 -- well before the first Moon landing in 1969. From 1966 to 1974, NASA's budget steadily declined. By 1974, NASA only had 54.8 percent of its 1966 budget. In dollars, NASA went from $5.93 billion in 1966, to $3.252 billion in 1974. Since 1974, NASA's budget seems to have consistently grown, specially since 1988. Unfortunately, the large budget increase from 1987 to 1988 did not reflect a new-found commitment to space by the government. The increase came from a lump-sum payment by Congress for a new shuttle orbiter to replace *Challenger*. From 1989 to 1991, NASA did enjoy a short period in which its budget grew in a very healthy manner. Since 1992, its budget has essentially stagnated or declined. After taking inflation into account, NASA's buying power has dropped precipitously.

However, Figure 2 portrays the NASA budget in a completely different manner. By presenting the agency's budget as a percentage of the total budget, readers may have a better appreciation for the US government's commitment to space exploration. There has been no significant US government commitment to space since the Apollo program. Following the funding peak in 1966 at just under 4.5 percent of the budget, the countries desire to explore space dropped dramatically to a record low of 0.75 percent of the federal budget in 1986. This low coincidentally occurred in the year of the *Challenger* tragedy. Since then, NASA reached another smaller peak in 1991, at about 1.05 percent of the federal budget. The bad news is, that this weak commitment to space is deteriorating further.

If someday the US takes space exploration seriously again, it will be necessary to increase NASA's budget to allow the agency to do its job. There are currently two issues that prevent a larger NASA budget. The first problem is the federal budget deficit and debt. Many of today's congressional representatives, especially Republicans, are very determined to balance the budget at all costs. The second problem is the manner in which Congress organizes its budget process.

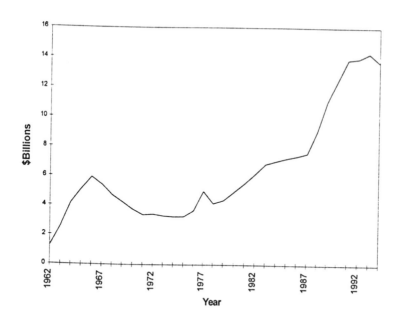

Figure 1. NASA's budget from 1962-1994.

Source: *Budget of the US Government, FY 1996.* p. 14. .Note: Fiscal Year 1977 included a transition quarter to adjust for a change in the time period of the fiscal year for the Federal Government.

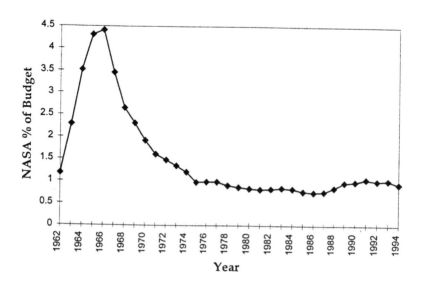

Figure 2. NASA's share of the US federal budget from 1962-1994.

Source: *Budget of the US Government, FY 1996.*

The House and Senate pass budget resolutions in the Spring and settle on an amount for the overall budget. The House and Senate Appropriations Committees then divide the budget for each of the existing appropriations subcommittees. Currently, there are 13 of these subcommittees in each house of Congress. NASA, as an independent agency, falls within the Department of Veterans Affairs/Housing and Urban Development/Independent Agencies (DVA/HUD/IA) appropriations subcommittee. The DVA/HUD/IA subcommittee, in both the House and Senate, is the catch-all for federal agencies that do not fit into any other federal department. The result is that a group of unrelated agencies with different missions and goals have to compete each year for a set amount of money. The heavyweights with the most power and money include the Department of Veterans Affairs (DVA) and Housing and Urban Development (HUD).

In fiscal year 1993, the DVA received $35.487 billion from an appropriations subcommittee that had less than $100 billion to spend. Next in line were HUD

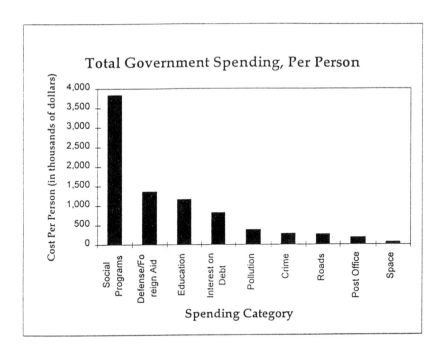

Figure 3. Total government spending per capita.
Source: *Budget of the US Government*, FY 1991.

agencies that include Army Cemeterial Expenses; the Consumer Product Safety Commission; the Neighborhood Reinvestment Corporation; and the National Commission on American Indian, Alaska native, and native Hawaiian housing. Keep in mind, that this only scratches the surface of the types of agencies in the DVA/HUD/IA part of the budget. The process concludes in late autumn, when a House/Senate conference committee convenes to resolve any differences in their individual budgets. This entire appropriations bill, which includes NASA, is then voted on in the House and Senate and then sent to the president for signature. Congress can, of course, take longer to get the job done. The 1996 fiscal budget was not resolved until April 1996, because of intense partisan fighting between congressional Republicans and President Bill Clinton. Clinton vetoed several funding bills in late 1995 and in early 1996 due to objections at cuts passed by Congress.

Spending on civilian space efforts also does not compare favorably with many other governmental functions. Considering spending by federal, state, and local governments in 1991, if one were to divide the amount of money spent on civil space activities among all US citizens, then the cost for each American would be $52 per year (see Figure 3).[50] By comparison, the per person cost of all current social programs and tax breaks costs each person $3,837 per year. Pollution control costs $385 per person, road expenses $255 per person, and post office expenses are $175 per person. When one examines all government spending in the US, it is quite apparent that space travel has not been near the top of the national agenda since the 1960s.

There is also the issue of civil space spending versus military space spending. By examining the budget authority approved for these items since 1969, it seems historically accurate to say that when civil space is in ascendancy, military space is in retreat. The reverse is also true (see Figure 4).

From 1969 to 1976, civil space clearly dominated military space because of the emphasis on Apollo. That trend reversed itself in the 1980s with the introduction of President Ronald Reagan's Strategic Defense Initiative (SDI) and a vastly increased military budget. The emphasis on military might also influenced the space arena as money was targeted at space-based missile defense research, more spy satellites, and the use of shuttles for dedicated military missions. More recently, the trend has been toward more equality in spending between military and civil space activities. It is hard to predict what will happen in the near future. A confluence of factors that include the budget deficit, reduced military spending, world events, politics, and future civil space policy will help shape things to come.

NASA and the US Economy

Since the beginning of the Space Age in 1957, attempts have been made to calculate the economic benefits that space activities have had on the American

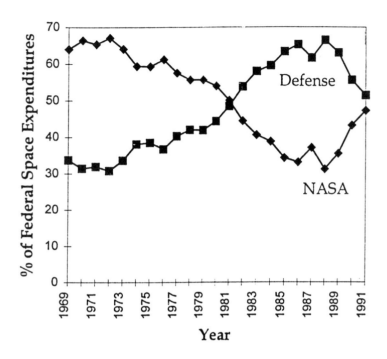

Figure 4. Comparison of NASA and defense space spending, 1969-1991.
Source: *Statistical Abstract of the US, 1993*, Table No. 995, p. 604.

economy. These studies are relatively few and are not widely known. However, the conclusions from some of these studies demonstrate that the impact of NASA spending is significant in relation to its overall size and budget. Whether it is jobs created, technology transferred or local economies bolstered, spending on the US civil space program has a positive impact across the nation. This economic aspect should not be under-emphasized.

As with spin-offs, it is imperative to realize, that our space program exists not only because it produces jobs, economic benefits or a good investment return. Everybody has their opinion, but I believe space exploration is important for two reasons: to explore the unknown and to help improve the lives of humans on Earth. Economic gains would generally fit into the second rationale, but they only benefit a small part of the global population. The real promise of space lies in the realization that the resources of the solar system can dramatically raise the standard of living of the poorest people on our planet. Imagine what cheap and limitless electricity could do to help people in the less developed countries of the world. This vision is possible through solar power satellites. Another incentive is the mineral wealth in the solar system's main asteroid belt that is sufficient to fuel human civilization for thousands of years. Mining asteroids in space could provide all the vital materials necessary to keep human economies moving right along. In addition, space-based manufacturing could help solve some environmental problems on Earth. The next chapter presents more of these concepts.

As for exploration rationales, which are unfairly trivialized by politicians and common citizens, the benefits are unpredictable. The desire to learn, explore, and to understand our celestial environment is an ingrained trait of many people. Imagine if (when?) the first fossil on Mars is discovered by an astronaut. Such event will forever change how humans view themselves and may significantly shake the very foundations of thought on human origins and place in the cosmos. Many of the belief systems held by people will get figuratively washed away by the revelation of this new knowledge.

The Midwest Research Institute (MRI) and Chase Econometrics Associates, conducted studies from a national viewpoint in the 1970s that examined how NASA research and development (R&D) spending affected the economy.[51] The MRI report followed the effect of NASA R&D funding on technology-induced gains in the US Gross National Product (GNP). The result showed that for every dollar spent by NASA on R&D, more than seven dollars were returned into the economy over an 18-year period.[52]

It is important to note that this 7:1 return ratio only applies to R&D dollars and not to every dollar spent by NASA. The MRI report showed that of the $25 billion spent on space R&D from 1959-1969, $52 billion was added to the GNP by 1970 and $181 billion by 1987.[53]

These impressive numbers prove that spending on government space activities can truly be considered an investment. Due to the huge size of the US economy, the space return-on-investment is not felt directly by average citizens. The money disperses throughout the nation. Just like spin-offs, the economic benefits of space are more subtle and invisible, but they are real. In a similar study, Chase Econometrics looked at the impact of NASA R&D funding, both from a short-term and long-term perspective.[54] In its short-term analysis, Chase used an economic model developed by the University of Maryland to show, that if $1 billion was transferred to NASA from other non-defense federal programs, 1975 manufacturing output would increase by 0.1 percent or $153 billion in 1971 dollars. Also, 20,000 additional manufacturing jobs would have been created in 1975. All these calculations assumed no change in the size of the federal government. In its long-term look, Chase calculated that US society's rate of return from NASA R&D spending was 43 percent. By contrast, the MRI study calculated a return rate of 33 percent. In considering all these statistics, it is important to remember that most scientific R&D will have a positive societal

TABLE 1
Estimated Sales and Jobs Created in Each State by Proposed FY 1990 NASA Procurement Expenditures (in millions of dollars)

State	Sales (in millions)	Jobs Created
Alabama	858.2	8,582
Alaska	35.5	236
Arizona	229.6	2,424
Arkansas	89.0	876
California	6,766.6	70,322
Colorado	490.3	5,381
Connecticut	601.1	6,224
Delaware	32.7	291
District of Columbia	75.2	990
Florida	1,297.50	14,756
Georgia	299.7	3,224
Hawaii	23.0	278
Idaho	21.4	242
Illinois	549.2	5,657
Indiana	348.3	3,253
Iowa	94.7	1,050
Kansas	172	1,697
Kentucky	142.4	1,358
Louisiana	535.0	4,583
Maine	36.0	386
Maryland	994.3	11,122
Massachusetts	382.3	4,208
Michigan	518.9	4,582
Minnesota	164.3	1,791
Mississippi	231.9	2,146
Missouri	342.3	3,427
Montana	18.7	180
Nebraska	47.7	566
Nevada	30.2	379
New Hampshire	58.7	626
New Jersey	506.0	5,411
New Mexico	135.5	1,242
New York	711.1	7,820
North Carolina	231.3	2,450
North Dakota	18.0	183
Ohio	928.7	8,545
Oklahoma	158.7	1,358
Oregon	67.3	731

The Case for Space

TABLE 1 *(continued)*
Estimated Sales and Jobs Created in Each State by Proposed FY 1990 NASA Procurement Expenditures (in millions of dollars)

State	Sales (in millions)	Jobs Created
Pennsylvania	602.2	5,955
Rhode Island	32.3	347
South Carolina	109.5	1,139
South Dakota	18.3	221
Tennessee	209.1	2,237
Texas	2,105.4	19,528
Utah	590.8	5,895
Vermont	21.1	226
Virginia	631.2	6,666
Washington	308.4	3,173
West Virginia	61.0	502
Wisconsin	193.0	1,991
Wyoming	28.0	210

Source: *The Private Sector Economic and Employment Benefits to the Nation and to each State of Proposed FY 1990 NASA Procurement Expenditures*, Management Information Systems, Inc., Washington, DC, 1989.

impact. Also, NASA R&D spending is not necessarily better than that of other research-oriented federal agencies.

Management Information Services of Washington, DC, released a comprehensive report in 1989, that examined the private sector benefits from NASA spending in fiscal year 1990.[55] The report concentrated on jobs created and how each US state was affected by NASA spending. The results produced both expected and unexpected findings, namely that "all states benefit economically from the space program..."[56] and "...that NASA procurement spending generates large numbers of jobs in industries not usually associated with the Space Program or the aerospace sector."[57] The study's aggregate figures are significant. Considering NASA procurement spending of $11.3 billion for goods and services, about 237,000 private industry jobs would be created. Also, companies would benefit from $23.2 billion in total industry sales, including $2.4 billion in corporate profits. Local, state and federal governments would also benefit from the generation of $7.4 billion in taxes (see Table 1)..

One important finding was that, as expected, many high technology and highly skilled jobs were created. Included in this group were many types of engineers (aerospace, electrical, civil, mechanical, etc.), mathematicians, chemists, aircraft engine mechanics, and other skilled labor. A more illustrative

TABLE 2

Jobs Created by Proposed FY 1990 NASA Procurement Expenditures Within Selected Occupations, Ranked by Relative Job Impact

Rank[a]	Occupation	Jobs Created
1	Aerospace Engineers	3,441
2	Mechanical Engineering Technicians	577
3	Electronic Repairers, Communication Eq.	915
4	Inspectors & Testers	1,556
5	Aircraft Engine Mechanics	881
6	Electrical Engineers	5,304
7	Mathematicians	123
8	Electrical Equipment Assemblers	2,047
9	Solderers & Brazers	395
10	Metallurgical Engineers	344
11	Industrial Engineers	2,288
12	Operations & Systems Researchers	1,359
13	Electrical Technicians	2,404
14	Mechanical Engineers	2,413
15	Grinding & Polishing Machine Operators	1,266
16	Metal Plating Machine Operators	394
17	Tool & Die Makers	1,272
18	Misc. Engineering Technicians	1,653
19	Computer Programmers	2,736
20	Marine Engineers	98
21	Purchasing Agents & Buyers	1,472
22	Technical Writers	329
23	Chemical Engineers	476
24	Computer Systems Analysts	1,620
25	Miscellaneous Engineers	986
26	Miscellaneous Science Technicians	344
27	Drafting Occupations	1,174
28	Civil Engineers	897
29	Mining Engineers	52
30	Chemists, except Biochemists	412

[a] Ranked on the basis of the percent job impact on the occupation.
Source: Management Information Services, Inc., 1989.

example is given in Table 2, which presents the jobs created in specific occupations and the relative impact of the NASA spending on each job. The top 10 occupations listed in Table 2 indicate that many non-space-related jobs are favorably affected by space spending. Examples of the non-space-related jobs include solderers and brazers, electrical equipment assemblers and mechanical engineering technicians. The rest of Table 2 includes occupations not normally associated with the space program: marine and mining engineers, technical writers and drafting occupations. This job information is very important to today's economic outlook. Why? The US economy is increasingly dominated by the predominantly low-wage service sector. This industry includes businesses ranging from restaurants to financial analysis that provide services to consumers. This industry does not produce goods; its success is based on the number of satisfied customers. The fastest growing occupations from now until 2005 are service-dominated and are conspicuously lean in high technology skills. As we approach the 21st century, the top job growth will include occupations such as the retail salesperson, registered nurses, cashiers, cooks, secretaries, and stock clerks.[58] Common characteristics among these and other jobs are their low pay, few benefits and long working hours. By comparison, many of the jobs for space exploration require high skill levels and therefore and earn much better wages. For example, salary offers in 1992 to students with a wide variety of engineering degrees ranged from $29,376-$40,679.[59] Other 1992 salary offers included $28,434 to math majors and $30,523 to computer science students.[60]

The MIS report also examined the impact of FY 1990 NASA spending on each US state. The basis for this analysis is the economic multiplier. By definition, this multiplier is the ratio of the total (direct and indirect) output requirements to direct output requirements generated from NASA spending. The best way to describe the multiplier is to use an analogy.

A direct output requirement can be likened to an aerospace company that NASA has selected to build a spacecraft. The lead company, called the prime contractor, will spend money to perform the job. The prime contractor is responsible for the overall spacecraft. Indirect output requirements can be viewed as the subcontractors who provide different parts of the spacecraft to the prime contractor for assembly. The multiplier ratio, total to direct output, indicates how much of an economic impact NASA spending has on states that do not directly receive NASA contracts. A few states receive most of the direct contract money from NASA. They include California, Utah, Texas, Louisiana, Alabama, Maryland, Virginia and Florida.[61]

These states are important beneficiaries of NASA money because they have major NASA field centers within their borders and provide home to many aerospace companies. More important, however, are the indirect benefits many non-space-related states accrue from subcontracts that help to fulfill a NASA program. It turns out that many midwestern and eastern US states benefit from space spending through subcontracts. Michigan, for example, has an economic multiplier ratio of 14:1.[62] The space program therefore helps Michigan economically through its companies that provide products to prime contractors.

Michigan is not known for its thriving space business, but the multiplier shows that it is more involved in space business than most people realize. Other states that benefit the most indirectly from NASA include Indiana, Kentucky, Arkansas, Oklahoma and Washington.[63]

Overall, this report clearly demonstrates the very positive economic influence of NASA spending across America. Space endeavor-related jobs can be considered favorably because of their emphasis on training and education, high wages as compared to many other occupations. Also important is the difference n motivation of employees who labor for a vision of human society beyond Earth, compared to those who work for an employer in other industries who receive most of the generated profits. It is good to know that the space program benefits distribute more evenly across the country than one might expect.

Spin-offs also have a positive impact on the economy. In Chapter 1, specific spin-off examples were introduced to show how widespread space technology has been transferred to private products. NASA estimates that more than 30,000 spin-offs have been modified for commercial use since 1958. However, NASA has not kept a comprehensive record of spin-offs and it would be impossible to track down each and every one. Despite this, the Chapman Research Group of Littleton, Colorado, released a report in 1989 that attempted to gauge the economic benefits of spin-off technology to the American economy.[64] This report used as its source the annual NASA *Spinoff* publication that gives examples of different kinds of spin-offs produced in the preceding year. The Chapman Research Group examined 259 spin-offs from 1978 to 1986 that were in the Spinoff document.

The report divided spin-offs into the different ways they were created. In the first group NASA technology was directly incorporating into a commercial product. In other cases, NASA employees commercialized a technology, or provided advice to companies about technology and hardware from NASA's Industrial Application Centers. In many instances, other government agencies exploited NASA technology, which helped to broaden the scope, and make it more attractive to the private sector. This list is not comprehensive, but shows that there is more than one way to move public technology into private applications.

This study examined 441 spin-offs, but it could only find sales or savings from 259 of the identified spin-offs. After separating these technology transfer products into nine categories, the study showed the derived sales and savings. These are summarized in Table 3. For example, industrial spin-offs produced company sales of more than $5.7 billion and savings of almost $68 million. This area produced the second largest monetary gain, after transportation, and is important because industrial spin-offs are technical and are not easily explained to the general public. The first chapter deliberately highlighted technologies that were applied to products that people could relate to.

TABLE 3
Benefits Realized from NASA-furnished Technology -- Case Applications
from *Spinoff* Reports. By Categories of End Use, Sales or Savings
(In thousands of dollars)

End Use Description	Number of Cases	Cases with Sales and Savings	Benefits Sales	Realized Savings	Total
Communications/ Data Processing	51	32	171,007	51,964	222,971
Energy	30	13	203,500	15,613	219,113
Industrial	170	107	5,767,649	67,837	5,835,486
Medical	61	31	2,003,036	30,613	2,033,649
Consumer Products	24	18	1,278,294	524	1,278,818
Public Safety	27	16	347,888	555	348,443
Transportation	40	18	9,887,865	116,623	10,004,488
Environmental	16	11	16,962	21,788	38,750
Other	22	13	1,654,989	10,232	1,665,221
Total	441	259	21,331,190	315,749	21,646,939

Estimates were obtained from company officials, or derived from company estimates of personnel or other types of savings. The 441 cases were reported in *Spinoff* magazine, 1978-1986; of these, 368 had acknowledged sales or savings, but 109 cases could not be estimated as to extent. Source: *An Exploration of Benefits from NASA Spinoff*, June 1989, by Richard L. Chapman, Loretta C. Lohman, and Marilyn J. Chapman.

It would have been pointless to list the technical improvements made to different types of machinery since the reader could only derive a vague appreciation of the technology application.

The total benefit from the 259 spin-offs is impressive: sales of $21.3 billion and savings of $315.7 million brings the total value of these spin-offs to $21.6 billion. If it had been possible to examine every spin-off, it can be assumed that the amount of money generated would have been significantly greater. Also, "discussions with corporate officials revealed 67 instances in which a product, process, or even an entire company would not have come into existence had it not been for the NASA-furnished technology. These represented 18% of all cases involving sales/savings and amounted to $5.1 billion in sales/savings."[65] All revenues from spin-offs brought in $356 million in corporate income taxes and created or saved 352,000 jobs, "[and] these jobs were in relatively high skilled categories."[66] This report furnishes further evidence that spending on the civil space program benefits people, companies, our governments, and the entire economy.

Commercial Space Industry

As alluded to earlier, private aerospace companies are the primary manufacture spacecraft and rockets. NASA can be thought of as the manager and overseer of the companies that build the space agency's hardware. The commercial space industry began to develop in the 1980s. Many new companies formed with the hope of offering products or services to NASA or of creating completely new industries. It is beyond the scope of this book to give a comprehensive survey of the commercial space industry since its inception. This subject could easily fill the pages of a book by itself. Therefore, after reading about the space industry and US government efforts to promote it, only four companies are highlighted, McDonnell Douglas, Lockheed Martin, Motorola and Orbital Sciences. These have been in the space business for varying lengths of time and who offer different products or services.

The future of this industry, which includes a wide assortment of firms, is hard to predict. Once launch costs drop below $1,000 per pound of payload, greatly expanded space operations could result. It could one day offer individuals the chance to leave planet Earth for orbit, the Moon, or more distant destinations.

According to the US Commerce Department, the commercial space industry expected to ring up sales of $6.49 billion in 1994.[67] This sales figure includes all major space activities that generate purely commercial revenues. They include Earth data-receiving stations, satellite services, commercial satellites, commercial launches, and remote sensing. Of these items, fixed and mobile satellite services were expected to reach sales of $2.3 billion. Satellite service sales grew by 230 percent from 1990-1993 and are expected to continue to grow in the future. Remote sensing, one of the smallest space markets with expected revenues of $300 million in 1994, grew by more than 160 percent from 1990-1993. It appears to have a very bright future. Also, the commercial launch market brought in $465 million in 1993 with estimated revenues of $580 million in 1994. This space segment roughly splits the international market with the line of *Ariane* rockets built and marketed by the European consortium, Arianespace. In addition, the Commerce Department reported that US companies manufacture about 69 percent of the commercial communications satellites in the world. In comparative economic terms, then-Deputy Commerce Secretary Rockwell A. Schnabel estimated that the export of US satellites and launch services in 1992 was "the equivalent of 60,000 imported cars."[68]

Since the early 1980s, the US government has attempted to promote commercial space development in America. One of the first attempts was the passage of the Commercial Space Launch Act of 1984. This legislation followed the lead of National Security Decision Directive 94, signed by President Ronald Reagan in May 1983, which announced US government support of expendable launch vehicle (ELV) operations. The congressional legislation had two objectives: to encourage rocket-builders to offer their launch vehicles for sale on the open international market and to make a single federal agency responsible for the licensing of commercial launches. In retrospect, this was a rather odd time to do this because US policy then dictated the end of the use of ELVs in favor of the space shuttle. By pursuing a space policy that dictated sole reliance on the shuttle, the federal government was at the same time encouraging ELV use and directing its demise. The actual start of the commercial launch market in the US began after the space shuttle *Challenger* accident in 1986. National Security Decision Directive 254 removed all commercial and foreign payloads from the shuttle in the aftermath of the tragedy.[69]

Henceforth, the shuttle was dedicated to only scientific and military purposes. The banning of commercial satellites from the shuttle is the main cause for the rebirth of the ELV industry. If the *Challenger* disaster had not occurred, it is likely that not a single US ELV would be in operation today. The other issue of the legislation, launch licensing, needed serious attention. President Reagan placed the Department of Transportation in charge of launch licensing. A company that previously wanted to launch its rocket had to weave its way through a figurative maze of requirements and regulations of many different federal entities. For example, after visiting the State Department to get an export license, a company would have to visit the Federal Aviation Administration to obtain permission to launch the rocket through US airspace. Another stop had to be made at the Coast Guard to assure that boat traffic would be restricted to at least three miles away from the launch. This only represented a few of the agencies that had to be consulted to get approval for a rocket launch. By making the US Department of Transportation in charge of launch licensing, companies can now do one-stop shopping with the government and avoid all the previous hassles of consulting different federal departments. This act was also amended in 1988 to cover the concerns about the insurance costs of the launch providers.

After the election of George Bush to the presidency in November 1988, his administration officially articulated its national space policy almost a year later on November 2, 1989. Among the Bush administration's top six general space priorities, commercial space activities figured prominently. Although the policy stated that no subsidies were allotted to the commercial space sector, other incentives were provided to help promote this sector of the American economy. Among these were the purchase of commercial space goods and services by the government, promotion of the US launch industry, and a focus on reducing the costs of space transportation. The Bush space policy also encouraged federal inter-agency cooperation to help promote the commercial space agenda, the streamlining of regulations in the industry, and the promotion of free trade in the worldwide space business. On February 12, 1991, Bush amended his space policy by issuing the US Commercial Space Policy Guidelines. The only significant addition to the president's space policy involved the concept of anchor-tenancy. In anchor-tenancy, the US government purchases goods and services from up-start space companies to help them get on their feet financially. Bush recognized, and the space industry knew for years, that reduction of spacecraft operational costs will provide the ultimate key to opening up the space frontier to businesses and individuals.

The International Launch Marketplace

Launching rockets into space is no longer the sole province of the United States and Russia. Today, rockets of varying size and capability are launched not only by the two space superpowers, but also by China, Europe, Japan, India, and even Israel. Commercially, however, the competitive participants include the US, Russia, Europe, China, and Japan. In the US, the big commercial rockets include the *Delta, Atlas,* and the *Titan* 4. The US government uses the *Titan* 4 primarily. The space shuttle has not been allowed to fly commercial payloads since 1986. Of the smaller US rockets, the relatively new *Lockheed Launch Vehicle* (LLV) and Orbital Science Corporation's *Pegasus* and *Taurus* rockets are available to commercial users.

In Europe, the *Ariane* rocket is the leader in the commercial launch business. Operated by the European Space Agency (ESA) and commercially marketed by Arianespace, the *Ariane* rocket has become ever more powerful since the launch of the first *Ariane* 1 in 1979. Currently, the *Ariane* 4 is ESA's primary booster and can place 9,800 pounds into geostationary transfer orbit (GTO). An even more powerful version, the *Ariane* 5, made its maiden flight in May 1996. Between these two boosters, Arianespace had a backlog of 36 satellites to launch as of March 1994.[70] In the 1980s, the *Ariane* and the space shuttle competed for satellite contracts and the two vehicles roughly split the market share equally. After *Challenger,* however, the *Ariane* rocket became the leader in the commercial launcher industry. Known for its excellence in service and operations, Arianespace will likely continue as the market leader throughout the 1990s, barring the development of new launch vehicles that are cheaper and more reliable to operate.

The Chinese have spent several years trying to market their *Long March* series of rockets to potential clients around the world. China has launched rockets since 1970 and has placed 32 satellites into orbit between 1970-1991. Attempts to offer *Long March* rockets for sale commercially did not commence until 1987. Marketed by the Great Wall Industrial Corporation, the *Long March* 2E is currently China's primary launcher that can place about 6,900 pounds into GTO. Although it launched its first commercial satellite in April 1990, China's space program has been disrupted by international events and diplomatic agreements. In the summer of 1989, the drama at Tiannenmen Square in Beijing unfolded before the world that pitted pro-democracy student demonstrators against the Chinese government. After the violent crackdown by the Chinese government and carried out by the army, the international community resoundingly condemned and sanctioned China for its actions. On June 5, 1989, President George Bush signed an executive order that banned the export of controlled items to China. It just happens that satellites are defined as controlled items by the US government and this subsequently prevented the export of two US-built Australian satellites to China. This ban stayed in effect until 1992 and was lifted only on the condition that China would abide by missile proliferation controls.

Since 1989, the US has been attempting to limit the number of commercial rocket launches made by the non-market economies of China and Russia. The US has signed separate agreements with China and Russia in the past few years to limit their influence on the international launch marketplace. The fear is that a flood of cheap launchers from Russia and China will undercut the business of Western launchers and cause some companies to go out of business. However, a possible benefit could come out of unrestricted launcher access by China and Russia. The expected downward pressure on launch prices might force companies in the western world to focus their attention on building new launch vehicles that are even cheaper to operate.

Japan is also developing its space launch capability and intents to become a competitor in the commercial launch market. On February 4, 1994, Japan launched its first H-2 rocket from Tanegashima Island. This rocket is made from Japanese-only technology and is a two-stage vehicle that can lift about two tons into geostationary earth orbit (GEO). Japan has been working on this rocket since 1984 and has spent about $2.5 billion on the effort. Although Japan also built the rocket for the nation's use, the commercial prospects for the rocket are questionable. Analysts do not think the H-2 will attract much business because of its high price despite the Japanese goal of making the H-2 profitable after 10 years of operation. Another potential problem for H-2 customers is that the rocket can only be launched during two 45-day periods each year.[71] Japanese fishermen, who are a powerful lobby in that country, are worried that the launches will scare offshore schools of fish.[72]

McDonnell Douglas Space Systems Company

The McDonnell Douglas Space Systems Company (MDSSC) is one of the most well known aerospace companies involved in the US space program. Operating as a part of the larger McDonnell Douglas Corporation, this subsidiary includes the McDonnell Douglas Delta Launch Vehicle division.

This division is responsible for all launchings of the company's *Delta* rocket. This rocket, like other expendable launch vehicles, is a modified missile that initially had the purpose of delivering nuclear warheads to the former Soviet Union in the event of war. Before the *Challenger* accident, *Delta* rockets were solely launched for the US government. After *Challenger*, and with the removal of commercial satellites from the shuttle manifest, aerospace companies, such as

McDonnell Douglas, had sufficient incentive to offer their rockets for sale to commercial customers. With that, the first commercial satellite launch from a US rocket took place on a *Delta* rocket in August 1989. The rocket successfully placed a British Satellite Broadcasting (BSB-R1) satellite into a 22,300 miles high geostationary earth orbit. From that first satellite delivery until May 1992, Delta rockets lifted nine more commercial satellites for a diverse array of customers like Indonesia, India, the North Atlantic Treaty Organization (NATO), and GE Americom. While in the process of competing for launches in the international marketplace, MDSSC also moved to upgrade the capabilities of its rockets. After launching its last *Delta* I rocket with an Indian *InSat-ID* satellite on June 12, 1990, MDSSC introduced its enhanced *Delta* 2 rocket that can be configured to handle a variety of payloads.

However, despite their foray into the commercial satellite launching business, this McDonnell Douglas subsidiary still counts on the US government for most of its business. "Government demand is the sole factor driving the design of the McDonnell Douglas medium expendable launch vehicle... Although some commercial sales have been made since 1987, loss of the government as the *Delta*'s main customer would derail the program."[73] This can be seen in the string of significant US government contracts MDSSC has won from NASA and the Defense Department. On January 21, 1987, the US Air Force awarded MDSSC a contract worth $316.5 million. MDSSC was to launch seven Navstar Global Positioning System (GPS) satellites that are used to locate a person anywhere on the Earth. In 1988, the Air Force exercised options for 13 more *Delta* launches of GPS satellites. It granted MDSSC a new contract in 1991 for five additional launches and the advance purchase of three more *Delta* 2's for launch in 1994. The first generation of GPS satellites are now in orbit and operating. The Air Force awarded MDSSC another contract on April 9, 1993, that is potentially worth more than $1 billion. It involves launching as many as 36 satellites for the second generation of GPS satellites to go into orbit beginning in 1996. Besides these defense-related contracts, MDSSC won a contract from NASA on November 14, 1991, to launch four science satellites with options for 11 more launches through 1997. It is important to know that the US government has a policy of launching its payloads only on US rockets. Thus, American rocket companies have a large, protected market. The danger in this is that there are no incentives to develop new launch vehicles that could cut the cost of putting payloads into space. In the current status quo, launch costs are very high, at about $10,000 per pound of payload. The US government is not providing any significant money to build new and possibly different spacecraft that could potentially reduce costs dramatically.

The Lockheed Martin Corporation

General Dynamics owned the space systems subsidiary of Lockheed Martin up to April 1994. Before the sale of this division to Martin Marietta and the subsequent merger of Martin Marietta and the Lockheed Corporation, General Dynamics encountered some difficulties in the commercial launch marketplace. Although it was in second place behind Arianespace in its market share for launch vehicles, rocket failures and financial losses finally reached a point at which the company decided to sell its launch vehicle assets. For Martin Marietta -- and now Lockheed Martin -- the acquisition was a boon. It transformed it into an aerospace company that now can build and launch satellites. Here is the story of the Atlas rocket since 1987.

In June 1987, General Dynamics made an initial commitment to build 18 *Atlas* rockets at a cost of $100 million. Not long thereafter, the company decided to expand its commercial launch business and build 62 additional *Atlas* vehicles for its hoped-for commercial customers.[74] The decision to sell expendable launch vehicles commercially resulted from the removal of commercial satellites from the shuttle. The company's first launch was on July 25, 1990, when an *Atlas* 1 carried the Combined Release and Radiation Effects Satellite (CRRES) into orbit for NASA and the US Air Force. However, even as their *Atlas* program built up momentum with this first launch, rumors were already circulating that General Dynamics wanted to sell its space division.

According to company information, General Dynamics Space Systems Division lost money in 1989, 1990, and 1991.[75] Although net sales rose during that time, operating losses persisted with red ink of $76 million in 1989, $395 million in 1990, and $68 million in 1991. These losses mounted even though General Dynamics worked its way to second place behind Arianespace in market share with a backlog of 35 satellites to launch at the end of 1991.

Many of the company's problems can be traced to three *Atlas* failures that occurred in less than two years. The first mishap took place on April 18, 1991. An *Atlas* 1 was destroyed in-flight when its hydrogen-oxygen fueled *Centaur* upper stage developed problems. Lost in the failure was a Japanese BS-3H direct-broadcast television satellite. One of the rocket's two *Centaur* engines failed to ignite properly and caused the launch failure. Internal and external investigation boards identified two possible causes for the failure.[76] Both investigating bodies thought it likely that foreign debris somehow made its way into the engine and caused the failure.

On August 22, 1992, an *Atlas* 1 failed to place a Hughes Communications *Galaxy* 1-R satellite into orbit because of another failure in the *Centaur* upper stage. This time, however, the cause of the problem was not linked to foreign debris. "Engineers believe both recent *Atlas* failures were caused by stuck valves that allowed nitrogen ice to form in the *Centaur* second-stage fuel pumps. In each case, only one of the two Pratt and Whitney-built engines ignited, causing the destruction of both rockets and their payloads."[77] A third major malfunction occurred when an *Atlas* placed a Navy UHF-1 satellite into the wrong orbit on March 25, 1993.

An *Atlas-Centaur* rocket begins its journey to earth orbit. (Courtesy General Dynamics)

General Dynamics soon decided to sell its space division and cut its losses. The company had invested $700 million into the *Atlas* system, but lost $618 million since 1989.[78] Rumors emerged in the second half of 1993 about negotiations between General Dynamics and Martin Marietta for the sale of the Atlas division.[79] By late spring of 1994, the sale had been completed and the *Atlas* belonged to Martin Marietta. Martin Marietta and the Lockheed Corporation later merged to become the largest aerospace company in the industry. This merger has epitomized the consolidation and acquisition trend in the space industry, as aerospace companies fight for survival in an era of defense and space-related budget cuts.

Four versions of the *Atlas* rocket are currently available for sale. These include the *Atlas* 1, *Atlas* 2, *Atlas* 2A, and the *Atlas* 2AS. The capability of each rocket varies; the *Atlas* 1 can place 4,950 pounds (2,245 kg) into a geostationary transfer orbit (GTO), while the *Atlas* 2AS can loft 7,700 pounds (3,500 kg). GTO is a temporary orbital position for a satellite before ground controllers circularize its orbit for placement into geostationary earth orbit at about 22,300 miles. Like McDonnell Douglas, Lockheed Martin has a rocket that is still used mainly by the US government. This is evident from the fact that the government accounted for 100 percent of *Atlas* sales in 1989 and 1990, and 66 and 70 percent in 1991 and 1992, respectively.[80]

Orbital Sciences Corporation

Unlike some of the big aerospace companies that just launch rockets, the Orbital Sciences Corporation (OSC) of Dulles, Virginia, can be likened to a full-service space company that provides several kinds of services to its customers. Founded in 1982, OSC has expanded its capabilities so that it can launch not only small satellites, but can also build and service a variety of satellites and upper stage boosters. This entrepreneurial endeavor is also preparing to work in the remote sensing and global communications industries.

Originally, the OSC founders wanted to develop an upper stage rocket for lifting satellites to higher orbits or sending probes on interplanetary trajectories after their initial launch into space by the shuttle or a rocket. Although company plans and goals diversified into other areas as the 1980s progressed, the *Transfer Orbit Stage* (TOS) has always been a product offered by OSC. Today, OSC has developed a family of seven TOS vehicles with varying capabilities that can meet many different needs. For example, the lightweight of the family -- the TOS 2LT -- can place 800 pounds into GTO and can be used on the *Delta* 2, *Titan* 2, and *Taurus* rockets. On the other hand, the most capable vehicle -- the TOS 21H -- can lift as much as 13,400 pounds into GTO, using the space shuttle. As of July 1994, the track record of the TOS includes two successful operations and no failures. Used for the first time in September 1992, a TOS successfully boosted the *Mars Observer* spacecraft in the direction of Mars by performing a 2.5 minute burn that accelerated the vehicle to a speed of seven miles per second. The other TOS success occurred in 1993 when NASA's *Advanced Communications Technology Satellite* (ACTS) entered geostationary earth orbit after a boost from a TOS.

As the 1990s approached, OSC saw its business, staffing, and revenues grow. From 1987 to 1991, OSC's revenues jumped from $25 million to $135 million and its personnel ranks grew from a mere 42 people to 869. In addition, OSC grew at a 150% annual compound rate from 1983 to 1991 and saw its backlog orders grow from $20 million in 1986 to $225 million in 1991. Orbital was on the move and the company began to set its sights on even more ambitious projects. The next major activity, the *Pegasus* air-launched rocket, kept the company busy for several years. OSC began to bear the fruits of their labors on April 5, 1990, when the first *Pegasus* rocket delivered the *Pegsat* satellite and an experimental US Navy communications satellite into a 320 nautical mile orbit. The launch took place without a hitch and the company was

A *Delta* 2 rocket lifts off. (Courtesy McDonnell Douglas Space Systems Co.)

ecstatic about its success. After the launch, David Thompson -- one of Orbital's founders -- reiterated that, "the success demonstrated today was not the result of luck. It doesn't happen that way. It was a result of many hundreds of thousands of engineering man-hours of work."[81] *Pegasus* is not a typical rocket. It is not even launched from the ground, but from under the wing of a large airplane. Three solid-fueled rocket stages power the *Pegasus* rocket and it can deliver from 400-1,000 pounds of payload to any inclination in low earth orbit. *Pegasus* can also be equipped with an optional liquid-fueled fourth stage to launch heavier payloads. Thanks to its unique air-launched design, the *Pegasus* can place twice as much payload into orbit compared to its ground-launched equivalent of equal capability.

Since that first successful launch, however, the *Pegasus* has had its share of problems. The second flight took place on July 17, 1991. A payload of seven 50-pound UHF communications satellites were to be placed in an 82° inclined orbit 389 nautical miles high. These Microsats did not reach their intended orbit because the first stage of the _Pegasus_ rocket did not separate properly from the second stage. Despite the problem and the ensuing investigation to find and fix the problem, OSC captured its first commercial customer for _Pegasus_. On October 21, 1991, OSC signed a contract with the Brazilian government for the launch of that country's SCD-1 remote sensing and data relay satellite. The SCD-1, launched successfully by _Pegasus_ on February 9, 1993, is the first of a two-satellite system that will collect environmental data from up to 500 automated ground stations. The Brazilian probe will also monitor deforestation and forest-clearing in the Amazon river basin. Several months later _Pegasus_ ferried ALEXIS into orbit. ALEXIS, which stands for the Array of Low-Energy X-Ray Imaging Sensors, took to the sky with _Pegasus_ on April 25, 1993. One of ALEXIS's four solar arrays was torn off during the launch. The final analysis determined that the Pegasus did not cause any of the problems.

During 1994 and 1995, more challenges arose for Orbital's air-launched rocket. Two failures of the four-stage _Pegasus_ XL has caused a backlog of science satellites that are still awaiting launch. Despite the failures, Orbital continues to win small satellite launch contracts and has a backlog of orders that need to be filled. OSC is also looking to develop an advanced _Pegasus_ that will help to cut the cost of launches. According to David Thompson, a next-generation reusable _Pegasus_ could be ready in three and one-half years. It uses advanced technology that could cut the cost of small satellite launches from $7,000-10,000 to $2,000 per pound.[82]

OSC also has experience in launching suborbital rockets other than the _Pegasus. Taurus_, which is a ground-launched rocket, developed by OSC and the Defense Advanced Research Projects Agency (DARPA), is a four-stage solid-fueled vehicle that is designed to be quickly assembled and launched. _Taurus_ successfully made its maiden voyage on March 13, 1994.

Beyond its launch activities, OSC is also trying to make a name for itself in the fields of space remote sensing and global communications. OSC's remote sensing work centered initially on its _SeaStar_ satellite. In March 1991, NASA awarded OSC a $43 million contract to build the _SeaStar_ satellite. In an innovative move, NASA devised the contract so that the space agency would buy at least $43 million worth of data from the satellite instead of providing the money to OSC up front. In this case, OSC is building the satellite with its money. The mission of SeaStar is to collect data on the levels of phytoplankton and chlorophyll in the oceans. NASA will use this data to study global warming.

In a bid to capture a share of the global communications market, OSC has proposed orbiting a constellation of 36 satellites that will provide two-way data services across the planet. This would include global messaging, emergency communications, search and rescue capability, remote asset monitoring, stolen automobile recovery, and cargo tracking. Called _OrbComm_, this global satellite system will be available to individuals, businesses, and governments anywhere

Iridium constellation of 66 low-earth orbiting satellites that provide global communication services. The satellites are located 420 nautical miles above earth. (Courtesy Motorola Co.)

on Earth. OSC plans to sell mobile and fixed terminals that can access the *OrbComm* network while also selling pocket-sized communicators so that individuals can access the system.

Motorola -- Iridium

Motorola is a familiar corporate name to many. It is one of America's top telecommunications companies. Although it had no previous stake in the space business, Motorola formed Iridium, Inc. Its ambitious effort is to build and deploy a global satellite-based communications system.

Iridium, as with Orbital Sciences' *OrbComm*, will be a constellation of satellites that provide worldwide communications services. The two systems offer different services. With *OrbComm*, a variety of data services will be available. People will be able to send and receive information through *OrbComm*. With Iridium, users get data and phone services. The appeal of Iridium is that users will be able to talk to anyone from any location on the planet.

Iridium consists of 66 satellites in LEO (in addition to seven spare satellites) that will provide instant communications anywhere on Earth. Six sets of 11 spacecraft positioned in different orbital planes will provide worldwide coverage. Users of the system will have either a pocket-sized hand-held telephone or a so-called Iridium Subscriber Unit to use for communications. Transmissions will either be relayed by satellite or will hook up with a local cellular system (with the use of dual-mode phones that can be used through satellite or cellular systems), if it is cheaper. Along with voice capabilities, Iridium will be able to transmit data, fax, and paging information. Work on this system is underway.

In August 1992, the Federal Communications Commission (FCC) issued Motorola an experimental license to construct and launch five satellites to demonstrate the system's feasibility. Motorola has also awarded the Lockheed Martin Missiles and Space Company a $700 million contract to build 125 spacecraft buses (the satellite frame and associated hardware). Motorola will build the communications electronics and antennas to each satellite. Lockheed Martin will provide most of the rest of the satellite components, including solar arrays, attitude control, propulsion, electrical power, and a host of other items. The Iridium system is expected to cost about $3.45 billion. Similar to *OrbComm*, this is another business venture that is truly global in nature. Along with US investors, Iridium is receiving investor money from companies and organizations in Canada, Venezuela, Saudi Arabia, Russia, Italy, Japan, China, and Thailand. Even the launching of the satellites will be an international venture. Currently, plans are for the US *Delta* 2 to launch 40 satellites, Russia's *Proton* to launch 21 spacecraft, and China's *Long March* 2E to orbit the remaining 12 satellites.

Chapter 4: Space Resources for Earthly Uses

"Human beings and the natural world
are on a collision course...."

Excerpt from a statement issued by
1,575 scientists, including 99 Nobel
Prize winners, in their *World
Scientists' Warning to Humanity*
paper of November 1992.

As our society approaches the 21st century, there is increasing awareness that the irresponsible activities of people around the world are adversely affecting the health of our beautiful planet. Consequently, environmental awareness is on the upswing in the US and globally. This can be seen, for example, by the Earth Day celebrations that take place every April and attract enormous interest and participation from people of all nationalities. However, the truly devastating environmental problems we now face are insidious and invisible. Ozone, the protector of life on Earth limits the amount of ultraviolet radiation that reaches the surface of our planet. Human-made chemicals have dispersed into the upper atmosphere and are steadily depleting the ozone layer, as everyone knows. Carbon dioxide, a seemingly harmless chemical compound in its own right, has been accumulating at ever-increasing rates in our atmosphere. Although there is no complete consensus in the scientific and non-scientific communities, some evidence points to an increase in the global temperature of our planet. If these trends continue, human society could face global problems that may extensively kill crops and food-bearing plants, raise ocean levels, flood coastal cities, and make summers and tropical locations even hotter than they are now.

The Ozone Layer -- Our Protective Blanket

What is ozone and how did it get into our atmosphere? Ozone consists of three oxygen atoms that together form a triatomic molecule (O_3). Where did the ozone come from? It is important to realize that the formation of the earth about

4.6 billion years ago did not create the life and scenery we see today. The evolution of a planet and the evolution of life took place over very large time scales. Ozone is generally considered to have formed in the Archean Eon. This time ranges from 2.5-3.8 billion years ago and coincides with the oldest known fossil that is 3.5 billion years old.[83] Before the ozone layer was generated, no life existed on this planet. Ultraviolet radiation bathed our new planet and no life crawled, swam, or slithered around for millions upon millions of years. Any galactic inhabitants visiting Earth then would have summarily dismissed this planet as unlivable. However, as time passed and the planet changed both chemically and physically, the conditions needed for biological life -- which include the ozone layer --fortuitously arose. From there on, life developed and flourished in our little corner of the galaxy.

The sun's ultraviolet radiation (UV) caused ozone formation. In our atmosphere, monoatomic oxygen (O) reacts quickly with other elements. It does not exist in its monoatomic form for very long. It is chemically more stable when it is a part of a compound. Therefore, the oxygen we breathe is diatomic oxygen (O_2, plus nitrogen and trace gases). As the diatomic oxygen is transported in the atmosphere, UV radiation reacts with and breaks down the O_2 molecules into individual oxygen atoms. The two individual oxygen atoms can then react with diatomic oxygen molecules to form ozone molecules. Ozone formation is a continuous process in the atmosphere.

The breakdown of ozone molecules by chlorofluorocarbons (CFCs) and other chemicals is almost a simple reversal of the formation process. The ozone depletion process begins when CFC molecules drift into the upper atmosphere and dissociate. The sun's UV radiation produces a reaction that results in the liberation of chlorine atoms from the breakdown of CFC molecules. The freed chlorine atoms are very prone to react with ozone molecules. For every free chlorine atom, thousands of ozone molecules can be destroyed.

Knowledge about the possibility of ozone depletion goes back to the early 1970s. The issue did, however, not receive serious attention until a British Antarctic Survey team released its results about the ozone layer in 1985. Their survey unexpectedly discovered that the thickness of the ozone layer over Halley Bay, Antarctica, had decreased by more than 40 percent from 1977 to 1984. This information immediately made ozone depletion an international issue.

Earth-orbiting satellites have been used for decades to monitor the health of the ozone layer. One of the first satellites to take this kind of measurement was

Opposite: *Nimbus-7* Total Ozone Mapping Spectrometer images show October monthly average ozone levels over a 12 year period. The color scale shows ozone levels. The ozone hole is evident from low ozone values (dark purple colors) apparent over Antarctica in recent years. (Courtesy NASA)

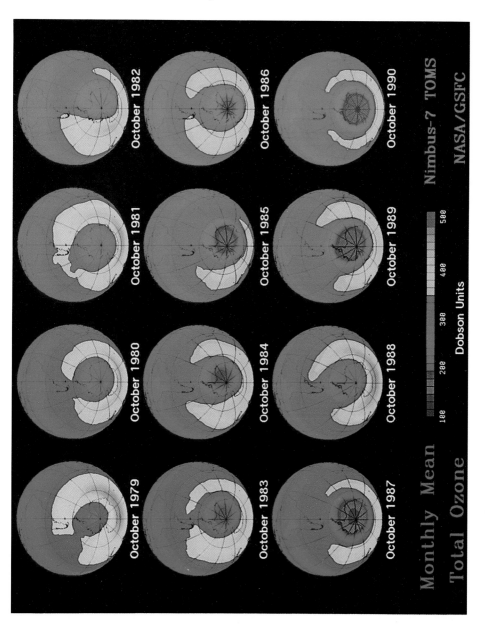

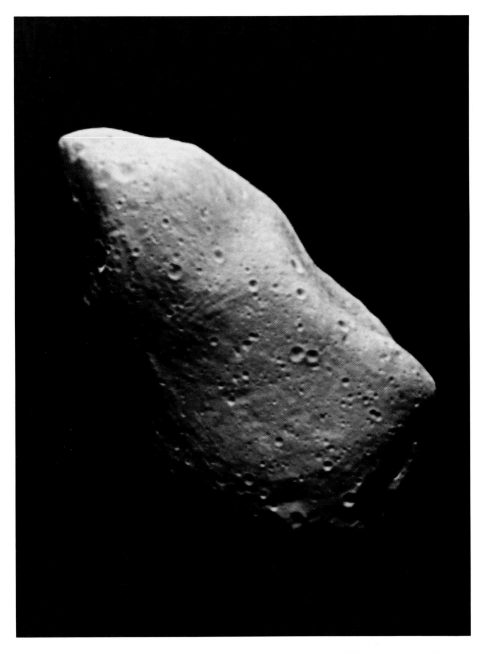

An image of the asteroid 951 *Gaspra* from NASA's *Galileo Jupiter* probe on October 29, 1991. Asteroids may not only be able to give us clues about the solar system's formation, but they also may have significant economic potential. See page 118 for discussion. (Courtesy NASA)

Nimbus 4. In the 1960s, NASA developed the *Nimbus* series of spacecraft designed to serve as test bed for a variety of new scientific instruments that were used on later environmental and weather satellites. *Nimbus* 4, launched on April 8, 1970, took ozone measurements over various parts of the Earth from 1970 to 1972. It found that the recorded ozone depletion was only half of what had been predicted earlier. Later in the decade came *Nimbus* 7, which took to the skies on October 24, 1978. Equipped with a Total Ozone Mapping Spectrometer (TOMS), the data received from Nimbus 7 were initially used to guide airliners around high concentrations of atmospheric ozone that could have harmed the passengers' health. In the 1980s, *Nimbus* 7 also contributed by mapping the global distribution of ozone and by monitoring the seasonal ozone hole that began to form over Antarctica. In the 1990s, the effort to understand and monitor the ozone phenomenon has accelerated with the launch of several spacecraft. On August 15, 1991, the former Soviet Union launched *Meteor*-3 that carried a TOMS instrument onboard.[84] This cooperative US-Russian mission continued ozone monitoring after *Nimbus* 7 ceased operating. Also, NASA now has two spacecraft in orbit that measure ozone levels. The *Upper Atmosphere Research Satellite* (UARS), launched by the space shuttle *Discovery* in September 1991, includes three instruments for monitoring ozone concentrations.

In September and October 1993, the TOMS on Russia's *Meteor*-3 recorded the lowest ozone levels ever measured over Antarctica.[85] Specifically, the TOMS measured an ozone level of less than 100 Dobson units. A Dobson unit is a one millimeter thick layer and is a measure of the ozone layer density. Ground and aloft data corroborated the space-based data. The US National Oceanic and Atmospheric Administration (NOAA) sent an instrument-carrying balloon into the atmosphere on October 6, 1993 that recorded an ozone level of only 90 Dobson units. The balloon also discovered that the ozone above Antarctica was destroyed at altitudes of 13.5-19 km. On the ground in Antarctica, a simultaneous measurement produced a reading of only 88 Dobson units. Continued ozone depletion could have numerous harmful effects for life on earth. Along with the possible widespread effect of disturbing the normal growth of plant and animal life due to radiation effects, the incidence of skin cancers among humans may increase.

Luckily, world leaders have taken the ozone threat seriously and have taken positive action to reduce the threat of future ozone depletion. In 1987, the United Nations passed the Montreal Protocol that called for all CFC-producing nations to cut CFC production in half by the year 2000. The US signed onto this treaty in 1988. Later accords included the London amendments in 1990 and the Copenhagen agreements in 1992. The London amendments specified the phase-out of CFC production by 2000 and the Copenhagen accord moved the phase-out timetable up to January 1, 1996.[86] The future health of the ozone layer, and the time it will take to regenerate itself, is currently being debated. Worldwide CFC production dropped from 1.26 million tons in 1988 to 820,000 tons in

1990.[87] The damaging effects from these chemicals will, however, be felt for years to come even after CFC production has ceased. Pessimistic estimates predict that even if CFC production ceased today, the ozone layer would not be fully replenished until 2060 at the earliest.[88]

Global Warming

The other major environmental problem is global warming. Carbon dioxide methane, nitrous oxides and CFCs are thought to heat the atmosphere. The burning of carbon-based fuels, such as oil and coal, introduces several of these gases into the atmosphere. The greenhouse gases trap the solar energy that reaches Earth. The atmosphere functions as a greenhouse to warm the planet sufficiently to sustain life, but there is the danger of an enhanced greenhouse effect. If carbon dioxide traps too much heat within Earth's atmosphere, global temperatures will increase. An extreme example is the planet Venus. The heat is absorbed in the atmosphere, but the planet radiates very little heat back into space. The planet is therefore so hot and forbidding. Severe global warming has long-term consequences, but is just as devastating to the future of life on Earth as ozone depletion.

What can be expected from global warming? Some predictions have global temperatures increasing in the next several decades. Some estimates suggest a temperature increase of 1.5°-4.5° C (3°-8° F). The magnitude of the problem will depend on how much the temperature increases. However, such relatively rapid global warmup will cause problems for humans and other animal species everywhere. Melting of the polar ice caps could raise ocean levels and flooding of low-lying coastal areas. The energy increase in the atmosphere could spawn storms that may be more destructive than any we have ever seen. Precipitation patterns may change and today's rich farming lands will be laid to waste. The major farming areas will likely migrate to the north where the temperatures will be more suitable for food crops. Deserts will expand, summers will be hotter, and life could get very hard for the people of planet Earth. "By the end of the decade [1990s] the effects [of global warming] will be obvious to the person on the street as well as to scientists", according to Anthony Del Genio, an atmospheric scientist at NASA's Goddard Institute for Space Studies. "If the models are anywhere near correct we will see continual increases in global temperatures. We'll see much more frequent droughts in many areas, hurricanes will be more intensive than any we've ever experienced."[89]

Through NASA's work and that of other scientific organizations, evidence is mounting that the amount of carbon dioxide in the atmosphere has been rising since the beginning of the Industrial Age. According to NASA research, there is evidence of a global temperature increase of 0.5°C (1°F) since 1850. This temperature rise has been caused by increases of atmospheric carbon dioxide concentrations of about 25 percent since the early 1800s. Climatologists at

NASA's Goddard Space Flight Center in Greenbelt, Maryland, suggest that carbon dioxide concentrations have increased 10 percent since 1958. Not only is the Earth's air temperature increasing, but sea surface temperatures are rising as well. Satellites have produced evidence that ocean surface temperatures have increased by 0.1°C annually between 1982 and 1988.[90]

How is it that so much more carbon dioxide is being dumped into the air? The gas is naturally regulated by earth's atmosphere. Carbon dioxide is, for example, exhaled by living creatures. As a counter-balance, plants consume it through photosynthesis and it is precipitated into certain rocks and into the oceans of the world. The problem is that human use of fossil fuels in vehicles and industrial plants has released so much carbon dioxide into the air that the earth's natural regulatory system has been disrupted. Carbon dioxide concentrations then increase and global warming is the result. "In the last 30 years [from 1961 to 1991] the CO_2 level has risen from 316 to 350 parts per million. Earth now has the highest concentration of CO_2 in the last 160,000 years."[91] The amount of carbon produced by our oil and coal-based civilization is truly staggering. "In 1988, at least six billion metric tons of carbon were added to the atmosphere -- about 5.5 billion tons from fossil fuel consumption and between 0.4-2.5 billion tons from burning or clearing forests."[92] As of 1991 the six hottest years on record were 1990, 1988, 1987, 1981, 1980, and 1986.[93,94] The relation to global warming is not firmly established. Using computer models at the NASA Goddard Institute for Space Studies, one estimate shows a 2°C temperature increase in the next 50 years due to greenhouse gases.

The greenhouse effect is directly related to the human consumption of fossil fuels. The production of carbon dioxide is causing an imbalance in the Earth's atmosphere. To mitigate the possible effects of global warming, energy alternatives need to be considered. In most discussions of global warming and energy alternatives, space-based solutions are not considered. This is either because no one knows about them, or our political leaders do not consider them as viable solutions. As a consequence of this rather limited thinking many innovative and revolutionary ideas do not reach the general public. Many people know of solar power, but few have ever heard of solar power satellites (SPS). However, before discussing a variety of SPS concepts, NASA's Earth Observing System (EOS) will be reviewed. This is NASA's contribution to the US Global Change Research Program; an effort by the US and other nations to learn how society is affecting Earth's environment and to determine the size of the threat we face from global warming. EOS is a mission for the present and SPS is a mission for the future. Both are important and informed citizens need to be aware of them.

The Earth Observing System (EOS)

The danger signals were flashing from the scientific community about global warming. A consensus emerged therefore in the 1980s, that one of

NASA's missions should study the earth to learn how the planet works as an integrated system. As a part of the overall program, it would study global climate change questions and analyze its findings for any answers. In 1987, former astronaut Dr. Sally K. Ride issued a broad-based report to then-NASA Administrator James Fletcher on possible future directions for the space program.[95] Among the options, Dr. Ride suggested an earth environmental mission to learn how our home planet functions. Such pressures led to the formation of the US Global Change Research effort and NASA's Mission to Planet Earth program. The national global change research effort is a wide-ranging, multi-agency activity that has a multi-billion dollar budget. Along with NASA, other participants in the program include the Departments of Energy, Defense, Agriculture, Commerce, and Interior. The Environmental Protection Agency and the Smithsonian Institution are also involved.

The Earth Observing System (EOS) is NASA's major contribution to the global change research project. Originally, the complete satellite and ground system was estimated to cost $30 billion by the year 2019. After spending this money, NASA was hoping to be able to answer some wide-ranging questions. How do clouds affect the energy balance on earth? How do all the forms of water affect the planet? How much carbon dioxide is being released into the atmosphere and how much carbon is being absorbed by the oceans? These and other extremely complex questions were the initial goals of NASA's EOS program. The EOS program first received funding from the Congress in fiscal year (FY) 1991 for the amount of $191 million.

However, even before the program took to the air debates ensued about how many satellites and instruments should make up the planned constellation. The debate was clear-cut, but the opposing sides differed by light-years. On one hand, several prominent experts thought that NASA should launch many medium-sized satellites to make all the necessary observations. In August 1990, 12 university scientists who had been appointed to a panel by the National Research Council issued a report entitled "The US Global Change Research Program". The panel recommended that NASA use smaller satellites in the second series of EOS probes while sticking with three large satellites for the first series. Then, NASA planned to build six large satellites that would weigh 25,000-30,000 pounds each. NASA could make multiple measurements of the earth over a 15-year period that would facilitate data analysis and continuous operations, by using large platforms packed with many different kinds of instruments. The merits of large versus small satellites became the focus of concern for EOS.

Even before the EOS program went into high gear, storm clouds loomed over the horizon that threatened NASA's plans. There were concerns about how to run the program. The outlook for NASA's and the overall US budget grew worse month by month and many politicians made rumblings that the scope of EOS needed to be restricted. Both the budget and program issues were scrutinized at the end of 1990 and into 1991. The satellite debate raged on into 1990, when NASA Associate Administrator for Space Science Lennard Fisk opposed the NRC report mentioned earlier. Fisk thought that demanding additional satellites would increase costs since it involved a new west coast launch site for polar-orbiting satellites. Furthermore, it required an improved Atlas rocket that had not yet been built.[96] All the while, advocates of the smaller satellite option argued that a launch failure of a rocket carrying one of NASA's big EOS platforms could cripple the program. A larger fleet of smaller satellites a launch failure would not set back the project so much.

In July 1991, in response to continuing concerns over EOS program cost and structure, an EOS External Engineering Review (EER) team formed to provide recommendations on how to reduce EOS cost and ensure its success. Led by Edward Frieman, a scientist with the Scripps Institution of Oceanography in La Jolla, California, the eight-member panel reviewed every detail of the program and issued its final report on September 30, 1991.[97] Although it continued to support its large satellite plan before the EER panel, NASA must have seen the writing on the wall. About a month before the panel released its report, NASA announced that it had decided to conduct EOS with 18 medium-sized satellites. The EER team came out with its report anyway and recommended the use of smaller satellites for the environmental observation mission. The panel's other views included, that NASA should use a smaller and cheaper rocket than the Titan 4 to launch EOS satellites, other US government agencies should be involved in the program, and the EOS Data and Information System (EOSDIS) should be reviewed. EOSDIS will be a computer system that will compile, process, and distribute EOS satellite data to scientists and other interested people around the world. In a press briefing on August 14, 1991, to explain NASA's change of attitude, Fisk stated that large EOS platforms would "give you the most measurements...in the least amount of time, for the least amount of money."[98] However, Fisk noted that a "serious drawback" to large platforms "is that they are relatively inflexible when technical problems occur or when budgets do not grow as much as expected."[99]

As all this took place, the US Congress began to cut back on projected EOS spending due to bigger concerns about the budget deficit and overall EOS program goals. The year 1991 saw Congress reduce the EOS program to what it thought was a more reasonable and affordable budget. Along with making EOS focus solely on global climate change issues, Congress mandated that NASA use smaller satellites and that the EOS budget would be reduced from $17 billion to $11 billion through FY 2000. In a final fit of budget-cutting in 1992, Congress decided that the amount of money for EOS through FY 2000 would only be $8 billion.

As now envisioned, EOS will consist of 18 medium-sized satellites to be launched from 1998 to 2012. Beyond that, other nations plan to launch satellites in support of the environmental mission. Japan, Canada, and the European

Space Agency (ESA) will either orbit their satellites or provide instruments to be placed on satellites launched by other nations. Japan, for example, is planning to launch its *Advanced Earth Observing System* (ADEOS) satellite in 1996. They will place ADEOS into an 800 km, sun-synchronous orbit that will be inclined 98.6° to the equator. This near-polar orbit will allow the satellite to conduct remote sensing of land surfaces, oceans, and the atmosphere. Japan also plans to build a follow-on satellite to ADEOS 1.

ESA plans to contribute several satellites through its Polar-Orbiting Earth Observation Mission (POEM). POEM is designed to be split into two parts. The first series of satellites, called *Envisat*, will be used for general environmental monitoring and study the chemistry of the atmosphere. The *Envisat* spacecraft are currently set to launch between mid-1998 and 2003. The other satellite series, called *Meteop*, will serve as weather satellites while also taking climatic measurements. *Meteop* is expected to be launched around 2000. Other international involvement includes Canadian contribution of a scientific instrument to be placed on NASA's first EOS satellite, EOS-AM1, that will measure pollution in the troposphere. The instrument will measure the concentrations of carbon monoxide and methane in the lower part of the atmosphere. The troposphere is that portion of the atmosphere between 7-10 miles in altitude. "Our goal is to see how these gases are distributed around the world", according to Ghassen Asrar, EOS program scientist at NASA Headquarters. "Currently, we don't have a consistent and systematic method of monitoring how the atmosphere moves greenhouse gases around."[100]

Another segment of the program involves a joint US and Japanese mission. The Tropical Rainfall Mapping Mission is set for launch in August 1997 aboard a Japanese H-2 rocket. The satellite will carry both US and Japanese instruments and will focus on tropical regions around the planet and measure rainfall amounts. One important question is whether tropical rainfall amounts are varying due to deforestation. Tropical rain forests not only harbor many varieties of plant and animal life, but produce a large amount of oxygen that we need to breathe. If tropical rainfall patterns begin to change and the forests start to die, problems arise for the global environment. That is why these missions are important. Very vital questions must be answered to fully understand how Earth is reacting to its human inhabitants. We need to determine how we can live on Earth and protect the environment at the same time. Without the space-based capabilities offered by the world's various national space programs, it is very unlikely that society will obtain a comprehensive understanding of Earth and its climate changes.

Although the EOS effort has been scaled back, the program is still large and promises to be a rich source of valuable scientific data. NASA is the agency principally responsible for all aspects of this program. NASA's EOS program consists of not just satellites, but also the data and information system that will handle all the information flowing from the satellites. EOSDIS comprises 60 percent of the program budget, with the other 40 percent devoted to the satellites.

Now that the design, size, and capability of the satellites are set, the space agency is diligently working to move the program from drafting boards to orbit. The plan is for NASA to launch six series of spacecraft with different missions and purposes. The series are called EOS-AM, EOS-PM, EOS-ALT, EOS-AERO, EOS-CHEM, and EOS-COLOR. There are three satellites launches in five-year intervals in each series, except for EOS-COLOR and EOS-AERO. Only one EOS-COLOR satellite will be launched, in 1998, and five EOS-AERO satellites will be launched in three-year intervals beginning in 2000.

The first satellite in the program, EOS-AM1, is set for launch in June 1998 aboard an *Atlas* 2AS rocket. This satellite includes five instruments that will study various aspects of the atmosphere and land. Besides the Canadian instrument to study the troposphere, the other four instruments will take radiation measurements, thermally track surface features, and image vegetation. The purpose of the Clouds and Earth's Radiant Energy System (CERES) instrument is to measure how much clouds and the Earth's surface reflect and emit radiation. As solar energy irradiates Earth, much of that sunlight is reflected back into space. The presence of clouds prevents some solar energy from reaching Earth's surface; a part of that solar energy is reflected back into space. At night, a clear sky allows radiation to dissipate back into space and cools the planet surface. Clouds also serve to trap heat near the surface and prevent it from rising back into space. CERES will try to determine the role clouds play in maintaining a consistent energy balance on Earth. Does more cloudiness prevent solar energy from coming in? Does a cloudier planet mean that the greenhouse effect will deteriorate? These are the type of questions scientists will try to answer.

Another EOS-AM1 instrument is the Moderate Resolution Imaging Spectro-Radiometer (MODIS). This device is able to measure changes in the amount of global vegetation present, record land and sea temperatures, and provide ocean color and land productivity images. It will monitor land and sea temperatures for evidence of global warming. MODIS will determine the health of different types of plant life from orbit. The two other instruments, the Multi-Angle Imaging Spectro-Radiometer (MISR) and the Advanced Spaceborne Thermal Emission and Reflecting Radiometer (ASTER) will also conduct important science. MISR will be equipped with nine cameras that will observe land, water, and plant features. ASTER's function is to visually and thermally track ground features on the surface of Earth.

The sister spacecraft of the EOS-AM series is the EOS-PM series. These satellites are to be placed in afternoon sun-synchronous orbits at an altitude of 705 km. They are initially set to be launched in December 2000. The three EOS-PM satellites will study cloud formation, precipitation, the amount of water and ice present worldwide, and air-sea fluxes of energy, carbon, and moisture. EOS-

AERO, will only carry one instrument and will be in an orbit with a 57° inclination and a 705 km altitude. SAGE III is the EOS-AERO instrument that will study the atmosphere located from the middle of the troposphere through the stratosphere to measure ozone, water vapor, aerosols, and clouds.

Plans call for another nation to build this satellite. EOS-ALT and EOS-CHEM, both set to reach orbit beginning in 2002, will do more work to solve the global warming riddle. EOS-ALT will be in a polar, sun-synchronous orbit and will conduct orbit tracking, determination and altimeter calibration. Its other goals are to measure the amount of sea ice and glacier coverage while measuring cloud heights and the vertical distributions of aerosols in the atmosphere. The mission of EOS-CHEM will be to measure the solar energy flux, solar ultraviolet radiation, atmospheric aerosols, ozone, water vapor, and trace gases.

The different measurements from these satellites are needed to verify global warming, determine its the major causes, and how we can make public policy to deal with the problem. Global warming, like other natural phenomena, is a complex event that involves a host of factors. It is unlikely that there is just one cause that leads to the planet's heating. It is more likely that a multitude of interacting causes produce global warming -- some of which humans may undoubtedly cause.

Solar Power Satellites

Solar power satellites (SPS) and other efforts to collect solar energy from space for use electrical power on Earth are based on a simple idea. Place huge arrays of solar cells into space, convert the solar energy into microwaves, and transmit that energy to Earth for conversion to electricity. Thus, if enough solar energy could be collected from space, human society would have at its disposal a clean, cheap, and virtually limitless source of energy to power its civilization. That is the goal. Our civilization is choking on its carbon-based fuel sources. These pollutants are the primary contributors to global warming. Solar power satellites constitute a possible solution to our global energy and environmental problems.

The SPS idea is three decades old. Back in 1968, Peter Glaser wrote an article in the *Science* in which he talked about the potential of solar-powered satellites. Since then, public and private research that has tried to quantify the requirements for such project has been very spotty. In the late 1970s, the US Department of Energy and NASA conducted a study to identify the basic prerequisites and possible problems associated with such an energy-producing system. The completion of the study lead to no further action because of the projected high cost of such a system. Despite this disappointing conclusion, private organizations, such as the Space Studies Institute are trying to fund research of possible SPS technology. However, very little activity is taking place beyond this. Other nations, e.g., Japan, are also conducting SPS research, but not at a level that will produce space-worthy technology soon.

Why is SPS an attractive option? First, an SPS system would produce no carbon-based pollution or contribute to the global warming problem. Many current electricity-producing power plants are fueled by such energy sources, e.g., coal. Therefore, even if the electricity produced is non-polluting, the coal used to generate this power is. Second, the electricity produced by a global SPS could be available to any area of the planet. The world's poorer nations would finally have access to cheap power and could develop their national resources much more quickly than if they had to rely on oil or natural gas. Third, the cost to use SPS technology -- once it has been perfected and demonstrated -- is potentially very low. In reality, the end-use (i.e., consumer) cost of SPS-generated technology will need to be competitively priced, if this idea ever has a chance of being implemented. Fourth, space-based systems can collect more solar energy than comparable ground-based solar systems. Solar power generation on the ground is hindered by clouds and precipitation, nighttime darkness, and atmospheric conditions like fog that block the incoming sunlight.

Solar collection in space can take place continually depending on the orbits of the satellites. In any future scenario, it will probably be wise to integrate SPS systems into a global energy plan. This should use several different types of non-polluting energy sources to satisfy the growing power needs of human civilization. In 50 years, we may all be driving electric cars, see oil used only in the production of plastics, and have our cities powered by a regional SPS system. Along with exploration, one of the best rationales for humanity's movement into space is to improve the standard of living of all its members. SPS is a concept that has the potential to radically improve life on Earth. Everyone needs to know about it and understand it.

To get a clearer image of SPS, I will discuss several different SPS proposals. Each study has its particular emphasis and goals. The various ideas the space community is examining will hopefully expand the reader's horizons. More studies have to be conducted, SPS technology requires more development, and possible SPS problems need to be resolved. However, it is time that the general public is introduced to this idea. What really matters is not what particular SPS system is ultimately used, but that our society uses its celestial backyard to enhance and protect its home planet.

The NASA/DOE Study

After NASA and the US Department of Energy finished their study of solar power satellites in the late 1970s, the Office of Technology Assessment (OTA) reviewed and critiqued it in 1981.[101] This section describes the OTA report's main elements and some of the agency's more important conclusions.

The NASA/DOE SPS study examined three types of systems: microwave, laser, and mirror transmission of energy. The microwave option received the most emphasis because the technology involved was more developed than that of the other options. A single satellite in the "reference system", as the microwave SPS option was called, included a satellite with 55 km^2 of solar cells and a 1 km diameter transmitting antenna located at one end of the satellite. This is definitely a huge structure for a single satellite. The large satellite size is necessary because solar cells are not very efficient at converting sunlight to electricity. A large number of these cells would be required to generate a substantial amount of power for use on earth. The satellite operates in a very straightforward manner. The solar cells convert sunlight to direct-current (DC) electricity that is then conducted to the circular transmitting antenna. Microwave transmitting tubes, called klystrons, convert the DC power into microwaves at a radio frequency of 2.45 GHz. The microwave energy is then sent to a particular location on earth where it is received by a large array of ground antennas. The energy is then converted back to electricity and the power is used in the local electrical grid.

With present-day space capabilities, the construction and operation of only one of these satellites would be an extremely impressive achievement. One SPS would weigh from 34,000-51,000 metric tons. Some 550-950 astronauts would have to build the mammoth structure. In addition, a complete reference system of 60 satellites that could each deliver 5 GigaWatts (GW) of power would be able to satisfy almost half of America's generating capacity in 1981.[102] To accomplish all this, we need a new heavy-lift launch vehicle, as well as new crew and cargo spacecraft. At the time of this report, no segment of the space community was vocally calling for the building of new and reusable launch vehicles that could be much cheaper to operate. More capable and cost-effective spacecraft would do much to aid the construction and operation of an SPS system.

Although no one discovered major technological obstacles in the investigation of the reference system, some environmental questions were raised. The effect of continuous microwave transmission through the atmosphere on people and the environment is a major concern. Another issue is the possibility of an SPS system causing interference with communications systems. The non-mandatory US standard for microwave radiation is 10 mW/cm^2.[103] An estimate of SPS exposure suggests that this may not be a problem. One study produced a result of 5 mW/cm^2 directly under the receiving antenna (rectenna) that converts the microwaves to electricity on the ground.[104] At the rectenna's edge, the exposure only was 0.1 mW/cm^2.[105] When operated at maximum power, however, the exposure at the rectenna center jumped to 23 mW/cm^2.[106] Although these exposure levels do not seem threatening, we need much more data to establish the effects of low level of microwave radiation on plants and animals. A new energy system is worthless, if it harms people in the process. Another possible concern is that microwaves might heat the atmosphere and cause local weather disturbances. Again, we need more studies to know for certain.

The second SPS option included the use of lasers to transmit the converted solar energy to the ground. The basic system is the same, requiring satellites and rectennas, but the satellites would be located in low-Earth orbit (LEO). The microwave option has the satellites in geosynchronous Earth orbit (GEO) at an altitude of 22,300 miles. In LEO, the laser satellites transmit the energy to low-mass mirrors in GEO that then reflect the laser beam to a receiving station on Earth. By using a laser in an SPS system, some advantages appear compared to the microwave option. Since they are very focused, laser beam diameters are much narrower than microwave beams. With microwaves, the diameter of the SPS transmitting antenna is one kilometer. Also, a hypothetical rectenna would have to be 174 km^2 on the ground. The huge land areas required for a microwave-powered SPS would require rectennas to be located in unpopulated regions or offshore in the open ocean. Health and environmental concerns might also make this necessary. With narrow laser beams, the overall mass needed in space for the satellites could be reduced. Rectennas could be smaller and closer to the intended electrical users. This would avoid the need to transport energy from distant locations. Other possible advantages of lasers include their better economy on a small scale, a greater redundancy and reliability of the satellites, and a lower expensive of a demonstration project. Another plus is that the environmental uncertainties of microwaves are avoided.

However, lasers also have their disadvantages. The first is that they are not developed to anywhere near the levels needed for this type of use. Microwave transmission is a reality. The experimental transmission of microwave energy began in 1964. Microwaves were also tested in 1974. Lasers do not perform well around clouds and other types of atmospheric phenomena that tend to disperse the beam. Infrared laser transmissions are absorbed by clouds and fog and this reduces the energy received by the rectenna. To resolve this problem, lasers might only be allowed to transmit in the absence of clouds or bad weather near the targeted rectenna. This, in turn, would require the satellites to be able to store a significant amount of energy until transmission to the ground becomes possible after the weather clears up. Another concern is the relative inefficiency of lasers in converting from one form of energy to another. Good conversion rates will be necessary in any SPS. If the conversion of laser energy to electrical energy is very inefficient, then their use could be ruled out altogether. At the rectenna sight, another issue to consider is the possible safety hazard a high-intensity laser beam presents. Unlike microwave transmission, lasers require much more development before they can even tested for use in an SPS system.

The final SPS type examined was the use of large mirrors in space to reflect sunlight to ground stations on Earth for conversion to electricity. As in the other options, this method had its advantages and disadvantages. In one proposal, called the "Solares Baseline", it was envisioned that 916 plane mirrors would constitute a global power system that could produce 810 GW at six energy conversion stations. The advantages of this concept is its simplicity of design and operation. No energy conversion takes place in space; the sunlight is transformed to electricity on the ground. No transmission equipment is needed, but a very large ground area is required to receive the sunlight. A large number of mirrors in space would increase redundancy in operations and their lightweight design would reduce transportation costs. As for lasers, the mirror option would not raise any environmental problems. A relatively inexpensive demonstration project could also position a few experimental mirrors into space. Many more studies are necessary to refine the mirror concept. It is currently as undeveloped as the laser SPS option.

It is important to remember that the various system configurations in this government study do not represent optimum designs. The report took a broad view of solar power satellites to give decision-makers the necessary information for an informed judgment about the next step. Unfortunately, no US government SPS work has been performed since 1981.

Reviewing solar power satellites and the current world energy situation, the OTA realized that an SPS system would not be ready by the time world oil supplies will dwindle. If work on a full-fledged SPS system began in 1981, a commercial SPS was not expected to be ready before 2005-2015 due to the huge nature of the project. OTA admitted that there were still too many open technical, economic, and environmental questions to start an SPS project in 1981. It recommended that a decision should be made on SPS before 2000. We are now approaching the year 2000 and no substantial SPS research has been performed. However, the idea is not dead just yet. Several organizations have released SPS proposals that vary from the initial NASA/DOE/OTA work. Furthermore, the SPS technology is maturing as new ideas emerge.

Project Phoenix

Using the environment as its main concern, the Illinois Institute of Technology in 1990 released a report entitled "Project Phoenix" that proposed conceptual solutions to the problems of global warming. Unlike the OTA report, Project Phoenix focused on preventing extensive global warming. Its authors thought that this global problem could be a major threat to humanity's future well-being. The following excerpt from the study illustrates their concerns. "At any but the best projections, the prospect is bleak, with great cost to humanity and the earth. If the worst fears of the scientists are right, *heroic actions may already be called for*. In any case, we cannot afford to wait and see."[107]

This two-part project is the result of senior undergraduate and graduate college work in the Institute's Institute of Design in Chicago, Illinois. Twelve students worked on this project in 1988 as a part of their course in the Systems and Systematic Design Class. After almost 10,000 hours of labor, they finally produced their two-part report. One part, called "Fire Replaced", used solar power satellites to correct the carbon imbalance in Earth's atmosphere. We are going to examine this, starting with the project's premises. As stated earlier, the authors of Project Phoenix felt that action was called for to prevent global warming. Although carbon dioxide was identified as the chief global warming culprit, other gases such as methane, nitrous oxide, and CFCs were also acknowledged as contributors. The authors then identified the specific problem as the carbon imbalance on Earth caused by human activities. Project Phoenix found that an annual carbon excess of three billion metric tons is being added to Earth's atmosphere. This occurs despite natural Earth processes, such as photosynthesis and the absorption of carbon into the oceans, that continually removes this element from the air. Although 204 billion metric tons of carbon are taken out of the atmosphere annually, 207 billion metric tons of carbon are being pumped into the air. The Project Phoenix team concluded that the surplus carbon is coming from human use of fossil fuels and deforestation that robs Earth's natural ability to remove carbon from the air. The goal, therefore, is to reduce humanity's dependence on fossil fuels by changing energy sources.

The SPS system used resembles the microwave option used in the NASA/DOE study. Solar cells, with an assumed 30 percent efficiency, are placed on many very large satellites in space to continually collect solar power. After converting the solar energy to microwaves, the energy is transmitted to rectennas on Earth. The next problem was to determine how much solar energy and how many satellites are needed to bring Earth's carbon cycle back into balance. If the goal is to supply all the worldwide energy in 2030, we need 116 ($\pm$) 13 billion m^2 or 103,000-129,000 km^2 of solar paneling. Due to the low conversion efficiency of solar cells and the energy conversion losses, one needs many solar cells in space using current technology. About 1,000 satellites are required to carry all these solar cells, if each satellite has a surface area of 116 km^2. If the goal is only to bring atmospheric carbon dioxide levels back into balance, estimates produced a somewhat reduced need for 70,621 km^2 of solar paneling. A total of 3,065 solar-powered satellites would be needed, if each spacecraft generated 10 GW of power from 23 km^2 of solar paneling.

To accomplish this mammoth project, the US and other spacefaring nations would need space-based capabilities currently unavailable. Also, Project Phoenix proposed achieving its goal in a much different way than the NASA/DOE study. In the NASA/DOE study and the subsequent OTA analysis, it was assumed that all SPS satellites and their components would be launched from Earth's surface. By comparison, Project Phoenix members decided that it would be more cost-effective to build the satellites from lunar ores and to construct them in lunar orbit. This presupposes capabilities that no country is ready to commit to yet. One obvious need is an extensive lunar base that can mine large quantities of ores and a continuous, year-round human presence. Fortunately, the Moon contains many valuable resources that we can use to build and operate a lunar base and to build solar power satellites. Along with oxygen, there is a significant amount of silica and aluminum in the lunar soil. That is where the proverbial lunar gold mine is located. By processing the ores in the lunar soil, liquid oxygen can be recovered and used in a lunar base life support system, as well as for rocket propellant for ships operating in the Earth-Moon system. What is more important, though, will be the ability to extract silicon from the lunar soil for use in making large quantities of solar cells. Since early in the Space Age, silicon-based solar cells have powered satellites and spacecraft operating within the orbit of Mars. This technology is well known and poses few risks. Along with oxygen and silicon, aluminum is another element bound up in ores in the lunar dirt. This and other extractable materials on the Moon could be used to build the structure of the solar power satellites. By using the local resources available, Project Phoenix estimated that the cost of building an SPS system with lunar materials is only one-twelfth of SPS satellites made from Earthly materials and lifted out of the planet's deep gravity well.

Another requirement for accomplishing Project Phoenix is a fleet of piloted and unpiloted spacecraft that can routinely operate in the Earth-Moon system. Currently, no nation on Earth can move people beyond LEO. Once the transportation needs are met, it is envisioned that Space Station Freedom would be used as a stop-off point between LEO, GEO, and the Moon.[108]

If all these needs are realized, the processed elements are launched in bulk quantities to an automated factory operating in lunar orbit. The method of propulsion for the raw materials is something called a mass driver. A mass driver is an electromagnetic device that can sling great quantities of material into space at low cost. Small-scale models of mass drivers have been tested on Earth, but we need more research to get to the real thing. Once launched into orbit, the materials are then caught at the lunar factory by the functional equivalent of a very large baseball glove. No attempt has yet been made to capture debris that is moving at several miles per second, but this is just a conceptual study. Once safely at the factory, the appropriate metals are mixed in a solar furnace to produce the frame and structure of each satellite. The lunar orbital factory will also be responsible for growing silicon crystals for use in solar cells. Few people will be required to oversee the work at the factory because it will be automated. Scheduled maintenance would be conducted and ships would frequently stop by to pick up their cargo. The final step in this scenario is to assemble each satellite near the factory. Once finished, lunar shuttles will move each SPS to its operational location high above Earth.

That, in essence, is Project Phoenix. By declaring an imminent environmental emergency on planet Earth, it advocates an SPS system regardless of cost. The project provided no cost estimates based on the probable assumption that it will cost less to build and operate a successful SPS network than to pay for the costs incurred from global warming. Although cost was not an issue in their report, the Project Phoenix authors were wise to look for ways to make the project more affordable. They met that goal by looking to the Moon. While reading the rest of the studies in this chapter, the use of lunar resources becomes more of the rule and than the exception. Future space exploration efforts will quite extensively depend on the use of the solar system's resources. Living off the land is a time-honored exploration tradition and it is one that will have to continue as people begin to migrate into space.

The Space Solar Power Program

Another entry in our look at solar power satellites is a study issued by the graduate students who attended the summer session at the International Space University in August 1992. ISU, as it is known, is a relatively new educational institution. It is an accredited university that offers a master's degree program for students who are truly motivated to work in the space industry. ISU will be discussed more fully in Chapter 5, but the focus of the learning is for students to acquire interdisciplinary experience in a variety of space-related fields. During the 1992 summer session, the students worked to create a development program for a commercial SPS system.

Unlike Project Phoenix, the ISU Space Solar Power Program took a broad view of the many issues involved with a project of this dimension. Along with technical and financial analyses, topics such as space law were reviewed to see how an SPS system might affect accords like the 1967 Outer Space Treaty. It examined future energy demands and population estimates to discover the potential demand for SPS and to determine the most probable users. After finishing all the background work, the ISU team came up with a step-by-step development program to test SPS technology that would eventually lead to an operational system. For anyone seriously interested in learning about the many issues confronting SPS development, this report is required reading.

To discuss whether SPS could be financially feasible, the team needed estimates about worldwide energy use and project where the largest growth would occur. Today, the largest energy amount is used by the industrialized nations of the world. For example, members of the Organization for Economic Cooperation and Development (OECD) gobble up about 48 percent of the global primary energy production.[109]

It is fairly well known, that the United States consumes more energy than any other nation, but how exactly do Americans use that energy? In 1991, 42 percent of American energy use was for transportation, 38 percent for industrial use, 12 percent for use in households and other dwellings, and 8 percent for commercial use.[110] Energy use in transportation systems will probably continue to grow. The ISU study noted that "one-half of the world's oil is consumed by

500 million road vehicles. Currently, the number of vehicles is growing faster than the human population."[111] Conversely, the less-developed countries (LDC) of the world only use about 16 percent of the world energy supply. A look to the future shows that a majority of the increased energy demand will come from those nations that are trying to become industrialized. "If the present [energy consumption] trend continues, in about 25 years the LDC's energy consumption will reach the present state of consumption in the industrialized countries. Assuming that the population growth is mainly located in these regions, a great increase in energy demand will occur."[112]

The ISU study examined a few models that attempted to predict future worldwide energy demands. The Lomer model suggested three scenarios that world energy demands could grow from two and one-half to five times that of today's consumption.[113] The Oak Ridge model predicted an energy shortfall based on the need to reduce fossil fuel consumption to prevent global warming.[114] The energy shortfall is based on the difference between fossil fuel demand and use.

In the population department, the ISU team examined estimates by the United Nations and the World Bank. In that study, global population estimates for 2020 ranged from 5.5-10 billion people based on three assumptions.[115] These assumptions included a 3-4 percent growth rate in LDCs, a doubling time of 20-40 years in LDCs, and a growth rate of less than one percent in developed countries. We should keep in mind that the current world population of five billion people is already straining Earth's environment and natural resources. If anything, the best prescription for what ails Earth is for fewer humans to be alive -- or else a migration from the planet. A negative population growth rate, however, is very unlikely anytime soon. Society needs to find ways to balance its needs with the those of the environment and other living creatures of the planet.

With all this information, the ISU students projected that space solar power has a potential market that could range from $77-$230 billion. A substantial portion of this market would go to developing countries that need more energy while they are growing. Based on this potential market, a five-stage developmental program emerged. It would incrementally test a variety of space solar power systems culminating in the construction and launch of the first operational 5 GigaWatt SPS in 2036. Results from the team's in-depth look at SPS technology, however, demonstrated a need for more research to be conducted before certain parts of the testing program could be pursued.

In the first program phase, they planned a near-term Earth-to-space beaming demonstration that uses existing facilities at relatively little expense. Plans call for a beaming demonstration to take place between the Arecibo radio telescope in Puerto Rico and a simple receiving satellite in orbit. The Arecibo facility, with its 300 m diameter dish that snugly fits inside a small valley, would serve as the transmitter. The satellite would be stationed in a 785 km orbit and would consist of a 10 m diameter collection dish and a rectenna to receive the microwave energy from Arecibo. One purpose of this demonstration is to show that useful amounts of microwave power can be transmitted and received over long distances. Depending on the radio frequency used at Arecibo, 430 MHz or 2.38 GHz, the amount of power received at the satellite would only be 10 or 60 Watts, respectively. If the satellite receiver had a 50 Watt light bulb on board, the incoming power might light it. In its schedule, the ISU team expected this program phase to cost less than $10 million and to take place over a five-year period.

The second step in this program is to show that microwave power can be effectively transmitted between two spacecraft. Drawing on its international emphasis, the ISU team set up a test that would occur between Russia's *Mir* space station and a Russian *Progress* spacecraft that is used to bring supplies to *Mir*. This $80 million mission will test the transmission of higher power levels than the first test and will establish the feasibility of powering an orbital space station with microwaves instead of solar cells. The mission would begin with the launch of a Russian *Soyuz* rocket to place the *Progress* transport into a 320 km orbit and permit it to dock to *Mir*. Along with the normal delivery of supplies, the *Progress* will bring the microwave transmission gear that will be used during the test. After unloading, the Progress will undock and move a safe distance away from *Mir* in preparation for the demonstration. Meanwhile, cosmonauts will attach the transmitter to the outside of their space station. The only thing then left to do is for the *Progress* vehicle to automatically deploy its rectenna. Once everything is in place, the test can begin.

The team estimated that 184 Watts of power could be delivered between *Mir* and *Progress*. Once this mission has been completed, the remaining goals in the Space Solar Power Program will focus on transmitting ever-larger power levels from space to Earth. This type of real-life incremental testing needs to take place if an SPS system is ever going to come to fruition.

In the first attempt to beam power directly from space to Earth, the ISU team envisioned a 10-15 ton satellite located in a 1,000 km high sun-synchronous orbit that is inclined 100° to Earth's equator. This is a near-polar orbit that theoretically would allow the satellite to beam energy to almost any location on Earth's surface. The plan calls for the SPS to beam its microwaves to a rectenna complex in Antarctica. There is no doubt that this third part of the project is at least a magnitude more complex and costly than the previous two tests. The previous microwave beaming tests are important, but the real work will occur when microwaves are sent from space to Earth. This test will allow engineers to see the real-time data concerning the ability, efficiency, and cost-effectiveness of power beaming. It will also answer other equally important environmental questions. Does extended low-level exposure to microwaves harm plants and animals? Do microwaves appreciably heat the atmosphere? What radio frequency is the best to transmit microwaves while also being economically efficient? These and other pressing questions must be answered in a thorough testing program before the world commits to an operational system.

If this SPS existed today, it would probably be very easy to see the satellite moving across the sky on a clear night. Sky-watchers would have to be within the orbital path of the satellite to see it, but the odds are good that it would be the biggest and brightest object in the night sky. The SPS would be visible mainly from the sunlight reflected off its 1,000 m^2 of solar cells. Add the 100 m^2 phased array transmitting antenna to the satellite and there is a very large spacecraft orbiting Earth. Plans call for launch of this satellite by Russia's *Energia* heavy-lift rocket from the Baikonur Cosmodrome in Kazakhstan. This is the site where most of Russia's large rockets are launched and is the only site where cosmonauts take to the sky. During the design process, the size of the SPS itself became constrained by *Energia*'s fairing. The fairing is the very top part of a rocket that protects and covers the payload during launch.

With a satellite of this size, a much greater level of energy will have to be generated. This SPS test will beam 150 kW of energy from its transmitter. Once the microwaves reach the one-square-kilometer rectenna in Antarctica, the converted power is estimated to be 30-60 kW. This third program phase had approximately $800 million cost over 10 years. Analysis showed that this project required more money and technology development. Subsequently, this SPS was to be used for scientific study. Recommendations from the ISU team include an emphasis on building more powerful transmitters, more sensitive rectennae, and work on using 35 GHz as the power beaming frequency. Also, more earth-to-space and space-to-space experiments were suggested to gain more experience with this type of technology. When and if this stage of SPS development is achieved, the next step -- a pre-operational system -- will be another magnitude more complex and capable.

Once space solar power beaming has been verified as safe and economical, one needs larger satellites with more solar cells that are capable of generating power to our energy-hungry civilization. The ISU study examined a few advanced, pre-commercial SPS satellites, but only one will be reviewed here. In any event, projects of this potential size require international cooperation and collaboration. Space development is going to become a truly global activity and the building of ISU's 1 MW SPS demonstrator may be the type of mission that truly begins to bring the world together. This SPS demonstrator is envisioned as being a tightly coordinated effort between the US and Russia. Estimates have the 65-ton SPS being built in a 10-15 year period at a cost of less than $1 billion. After lifting the SPS parts into low earth orbit with the use of Russia's *Energia* rocket, the satellite would be assembled in space by shuttle astronauts. Upon completion, the SPS would use its thrusters to guide itself to a final orbit 20,309 km high. At that altitude, it will take the satellite 12 hours to orbit Earth. On each orbit, the SPS will be able to transmit from its 35 GHz antenna for 6.4 hours. This satellite will require a solar array area of 8,903 m^2. With this option, gallium arsenide solar cells were chosen over the traditional silicon cells. Gallium arsenide cells are more efficient at converting solar energy to electricity than silicon cells, but their cost of production is currently more expensive than their silicon counterparts.

If placed into orbit in the near-future, the ISU team estimated the 1 megawatt SPS to cost $13.7 billion. Long-term costs, assuming adequate technological advancements, drop down to $691 million. Once accomplished, the goal of a limitless supply of clean energy will not be far from reality. Much more research needs to be done and this report offers but one way to get to SPS. However, the work of this dedicated group of students in Kitakyushu, Japan, shows that SPS can transition from blueprint to hardware, if world leaders would only realize the potential benefits of this energy alternative.

The Lunar Power System

All the space-based solar power plans reviewed thus far are premised on the notion of using large structures in space to collect this energy and then transfer it to Earth for use. Some proposals use lunar materials for construction, some plans use different transmission frequencies, and some missions are justified for environmental reasons. However, other rationales are more broadly-based. Is an orbital SPS system the only way to effectively use the Sun's energy for human activities? The answer is a resounding "No!", according to the creators of the Lunar Power System Coalition (LPSC). Instead of positioning satellites into space, the Lunar Power System (LPS) wants to place huge areas of solar collectors on the limbs of the Moon. The collectors could receive as much sunlight as possible. Unlike the previous proposals, the solar collectors for the LPS are firmly attached to the lunar surface. It is also likely that much of the LPS will be made out of lunar materials. There would be no need to place many large satellites around the Earth to get the job done. Constructing a LPS would undoubtedly involve a substantial permanent human presence on the Moon. However, considering that an LPS would be only one aspect of lunar activities, the investment in the Moon could be well worth the effort.

The method of operation of the LPS is similar to previous technologies. The solar energy is collected at the Moon, converted to microwaves, and then transmitted to Earth-based receiving stations. Once at Earth, the microwaves will be converted back to electricity for use. According to the LPSC, the LPS concept offers many environmental and financial incentives. They make it an attractive project compared to other currently used forms of energy. Some of the possible environmental benefits of construction of an LPS include a reduction of sulfur and nitrogen dioxides in the air, no carbon dioxide buildup in the atmosphere, a reduced need for nuclear fission plants, and improvements in food and water quality.[116] In addition, LPS may have advantages over Earth-based solar power stations. "The LPS concept shares many advantages with terrestrial solar power concepts, but has unique advantages. The intensity of sunlight is greater on the Moon, no clouds interfere, construction [materials] can be lightweight in low gravity and absence of winds, rain, hail, dust, earthquakes, etc., cleaning problems would not exist, and issues of energy storage during dark hours on Earth can be avoided."[117]

The LPSC is hopeful that its proposal will become cost-effective in the long-term. "Early calculations for a Lunar Power System predict that the long-term economics will be exceedingly robust, relative to present energy/power costs, without even considering [the] positive influence on the environment...."[118] Besides that, LPSC work predicts that the most economically uncertain part of LPS is with the rectennas on Earth and not the lunar facility.[119] Although it is difficult to determine the true cost and effectiveness of this project before delving into the engineering details, a generalized plan to pay for an LPS system has been suggested assuming a total cost of $1 trillion over a 10-year period. If the taxpayers of the US, the Commonwealth of Independent States, Europe, Japan, and other nations contribute, the cost for each of these projected 400 million taxpayers would only be $250 per year for 10 years.[120] The key to success is to make LPS an international effort in which the project is paid for, and used by, all the people of the world.

This concludes the discussion of space-based solar power. A variety of intelligent and thoughtful ideas have been proposed over the past 10 years, but they all currently share one thing. All the SPS/LPS concepts exist only as paper studies. Why? There are probably many reasons for the current state of affairs for space-based solar power. Some of the more important reasons include a lack of maturity in some SPS technology, insufficient public knowledge about SPS, and a fear by political leaders who do not seem to see the importance of investing in SPS research.

Sadly, the SPS situation is a microcosm of the space exploration dilemma in the 1990s. As for you, I hope that you now have some basic knowledge about some space-based solar power concepts and their potential. If you are really interested in the possibilities of SPS, you should read the actual reports reviewed here (for lists of organizations that you can contact to buy these documents, see **p. 183**). After becoming informed and excited about SPS and space activities, the next step is to become a vocal advocate.

That will have to wait until Chapter 5, because now we are going to briefly look at the exciting world of asteroids. These are not just chunks of rock floating in space. Quite to the contrary, this collection of planetary rubble left over from the solar system's formation contains treasures that can make inhabiting our celestial neighborhood much easier than previously thought.

Asteroids

Up to this point in the Space Age, asteroids have generally been considered to be the "black sheep" of the solar system family. They have not had the scientific sex appeal of our neighborly planets like Venus, Jupiter, and Saturn. Asteroids also seem to lack the mystery and intrigue of comets that have zoomed nearby Earth since before recorded time. Usually, asteroids are identified by the average person as rocks floating somewhere out in space. The most attention given to asteroids is when one of the rather large ones comes uncomfortably close to Earth. There is ample evidence of asteroid or cometary impacts on Earth. The Chixculub crater in the Yucatan serves as the most striking reminder that celestial bodies do hit the Earth. There is also the Manson crater in Iowa, Meteor Crater in Arizona, and evidence of a mid-air cometary explosion over Tunguska, Siberia, in 1908 that flattened acre upon acre of trees. Only recently have some world leaders and organizations started to take the threat of an asteroid or cometary impact on Earth seriously. It is scary to realize that all the reports of asteroids coming close to Earth were discovered at or after the passage of the object. There could be an undiscovered asteroid or comet on a direct collision course with Earth right now. We do not know if this is the case because no large-scale effort has tried to locate and identify all the asteroids of the solar system. New asteroids are always being discovered, but the rate of discovery is slow and the effort lacks sufficient scientific resources. The stakes are truly high in this matter. If a 5 km wide asteroid landed in the middle of Times Square in New York City, millions of people would die instantly. The resulting impact would throw up so much dust and debris into the atmosphere that sunlight could be blocked from Earth's surface for days, weeks, or months. A nuclear winter could result that could kill crops and plants of all types and throw human society into unparalleled misery for possibly generations. If the situation is bad enough, our species could vanish. Nobody knows for sure, but consequences of this type of impact would be enormous. The odds of this happening are supposedly very low, but the results of an impact would be catastrophic.

Fortunately, asteroids have a good aspect that has largely been ignored. Once human society begins piloted deep-space exploration, asteroids could be mined for the minerals and volatile chemicals that comprise these rocky bodies. We have to discuss first, however, where asteroids come from, what they are made of, and how they can help future space travel.

There are two general groups of asteroids in the solar system: the main-belt asteroids and the near-Earth asteroids. Located between the orbits of Mars and Jupiter, the main-belt asteroids are concentrated in specific orbits. It is possible that these remnants from the origin of the solar system are the pieces of a protoplanet that did not quite form. Astronomers have found many of these rocky bodies and precisely determined their orbits. Also, NASA's Jupiter-bound *Galileo* probe took the first photographs of two main-belt asteroids, *Gaspra* (*see* photo on page 98) and *Ida*, as it passed through the belt. The other general asteroid group, the near-Earth asteroids, consist of three subgroups of asteroids that are located in different orbits. Called the *Apollos, Atens,* and *Amors,* these asteroids are much smaller in number but much more significant in their possible utility by humans. The *Amors* do not quite make it to Earth, they get as close as 0.3 astronomical units (AU) to our planet, but they primarily cross the orbit of Mars. The *Atens* and *Apollos*, however, both cross Earth's orbit as they make their way through the solar system. These bodies probably offer the greatest potential for scientific study and, ultimately, mining. Two important determinants will be the finding of an asteroid with an ore composition that is rich in needed minerals and economical enough to mine.

Meteorites, on the other hand, are possibly pieces of asteroids or other debris in the solar system that have made their way to Earth after being captured by our planet's gravity. It is possible to suggest the compositions and types of asteroids in the solar system by examining the variety and composition of the meteorites found on Earth. In a broad sense, meteorites are classified into three categories. They are the stony, iron, and stony-iron meteorites. All three groups are further divided into more specific subclassifications. The meteorites of greatest interest are the carbonaceous chondrites and those with a large quantity of metal-bearing ores. A carbonaceous chondrite is a meteorite that is undifferentiated and primitive. When it formed billions of years ago, it did not experience any type of heating that would cause the various elements to separate and later recrystallize. With differentiation, the less dense elements rise to the surface, while the denser elements sink to the core. To give an example, the Earth has undergone differentiation since its formation about 4.6 billion years ago. Our planetary core consists of iron and nickel and Earth's surface contains less dense elements like silicon, carbon, and oxygen that are bound up in a variety of chemical compounds. Also, carbonaceous chondrites have been found to contain water-soluble salts, clays, organic matter, and about 10 percent chemically bound water. If this meteorite is subjected to heat, most or all the water and other light elements vaporize and escape. This is why few carbonaceous chondrites reach Earth's surface -- the heat of atmospheric entry causes this them to break up quickly. With this type of asteroid hopefully being fairly abundant, it is possible that water in the asteroid can be mined and separated by electrolysis into liquid hydrogen and liquid oxygen for use as rocket propellant. Thus, spacecraft will not have to carry a full load of fuel for an entire mission when departing Earth.

The other types of meteorites, the iron and stony-irons, all have some metal content of varying amounts. A group of the iron meteorites, called the *Ataxites*, are known to contain 40-60 percent nickel. In addition, some stony chondrites contain significant amounts of iron that may someday be useful in building Mars habitats, space stations, or space ships.

Asteroids are definitely more than just rocks. If properly utilized, this celestial rock pile can open up the space frontier to human society sooner than most people realize. It could advance the day when space travel becomes as routine as jet travel is today. To get an idea of how an asteroid might be used,

A view of the lunar north pole taken from NASA's *Galileo Jupiter* probe on December 7, 1992. The Moon's north pole is located just inside the shadow zone, about a third of the way from the top of the illuminated region. There is speculation that frozen water may exist at the lunar poles. (Courtesy NASA)

it is time to revisit the International Space University. Another group of graduate students devised an asteroid mission during their summer studies.

The International Asteroid Mission

ISU students who were spending a summer at York University in Canada in 1990 developed the International Asteroid Mission (IAM), a scenario to mine an asteroid. Working at the Institute for Space and Terrestrial Science, they proposed a plan that had the mission to mine water from Asteroid *Orpheus 3361* in as economical a fashion as possible. The mission would be to deliver the water to a space station in lunar orbit where the water would be processed into fuel and used by spacecraft in the Earth-Moon system. They had to consider several competing factors in deciding how to set up this mission. To keep costs low, mostly available technology would be utilized. The asteroid of choice had to be within a reasonable distance to keep round-trip mission times short. They also had to determine the amount of asteroid material to be mined and the market value of the rocket fuel once available for sale. After answering these and other questions, a hypothetical mission developed. This may not be how future missions will exactly be pursued. However, the report breaks new ground by demonstrating that the concept feasibility and its great benefit to future space efforts.

The mining mission would begin with the launch of a Russian *Energia* advanced lift vehicle from Cape York, Australia, on January 4, 2007. Even before, however, several robotic missions were deemed necessary to ascertain that the chosen asteroid had enough water to justify using it.

Starting with a fly-by mission designed to verify the asteroid's orbit and surface composition, more complex follow-on missions would completely map the body and shoot penetrating devices into its interior to better determine the asteroid's true composition. A decision could then be made on if to use that asteroid in the mission, by repeating this at several places on the asteroid and returning fragments of the asteroid to Earth for study. In this case, the IAM study team chose to mine *Orpheus* 3361. Orpheus is thought to be a carbonaceous chondrite and, therefore, a good candidate for this mission. Although the study noted that *Orpheus* may not be the best choice as more asteroids are subsequently discovered, it ultimately fit mission needs. Besides containing water, *Orpheus* orbits Earth in a manner that made it a prime choice. Its path crosses Earth's orbit and its perihelion -- the position in an object's orbit that brings it closest to the sun -- is 0.82 astronomical units (AU). Earth orbits 1 AU from the sun. At aphelion, the farthest distance from the sun in an orbit, *Orpheus* travels out to about 3.24 AU. Using a more familiar unit, the furthest distance between Earth and *Orpheus* is about 336 million kilometers. Other benefits offered by *Orpheus* are an orbit that is virtually in the plane of the ecliptic and it allows the IAM mission from 1.3-3.3 years to complete. *Orpheus* is 0.8 kilometers in diameter.

Now that the asteroid has been selected, the mission can start. Once the crew vehicle and the three cargo vehicles have been constructed in low Earth orbit, the crew of eight people can begin their trip. On the outbound leg of the flight, it will take 516 days to reach *Orpheus*. The propulsion system used is nuclear thermal propulsion (NTP). This system, tested in the late 1960s and early 1970s before being canceled by Congress, uses a nuclear fission reactor to heat the rocket propellants. It thrust these gases out of the rocket nozzle. NTP is more efficient than all types of chemical propulsion and will probably be necessary in future space exploration. In the first trip to *Orpheus*, the cargo vehicles are loaded with all the necessary machinery and mining equipment that will be needed to complete the job. Upon arrival, the NTP plant is removed from the ship and placed on the far side of the asteroid to take advantage of its shielding effect. The NTP will then provide electrical and thermal power for the mining operations. The next job is to unload the mining equipment and to set it all up at the operations site. Once that is accomplished, the crew vehicle will land a few hundred meters from the mining site.

With plans calling for a 275 day stay on the asteroid, it is estimated that 180 days will be devoted to actual mining. A typical day at *Orpheus* for the crew might involve the following routine. After getting a good night's sleep in the "home-away-from-home" ship, the astronauts prepare for the day ahead. The crew is divided into two work groups consisting of four people each. Each group works a four-hour shift each day and will have the daily goal of conducting two blasts that dislodge about 39 m^3 of rock per blast. With the smaller sun watching over them in the distance, the first shift of workers take their stations in the crew vehicle where the mining operations are coordinated and controlled. Most of the mining process is automated, so there is no need for the astronauts to leave the safety of their ship very often. Looking out of the ship's windows at the mining site, the crew views a cross-shaped piece of machinery that performs most of the work for them. Four square frames that surround a central square frame comprise the mining facility. The rocks will be processed in the middle section. On top of the frames are two mobile cabs designed to move across the top of the structure. One cab drills holes into the asteroid and plants the explosive charges. The other cab collects the rocks after the detonation and moves the rubble over to the processing facility. In case of any problems, the repair robot KiTi can be telerobotically controlled by the crew to fix any equipment problems. It may be that one person is in charge of the overall operation of each shift while the other three workers concentrate on their assigned tasks.

Passing through one cycle of operations on this conceptual mission, the first task would take place as Cab #1 starts drilling holes in the asteroid at a predetermined site. Cab #1 would then place explosives in each hole that will eventually be detonated to blast out the rock. This is the basis for the drill-and-blast method of mining. Each blast area covers 40 m^3 and uses 5.2 kg of explosives. After detonating the explosives, Cab #2 would collect the rubble in a large bag and move it over to the processing center. Once unloaded, Cabs #1 and #2 would begin preparations for the next blast while the products of the previous blast begin a two-part processing procedure. Pyrolysis is the first process that removes the chemically bound water from the rocks. The freed water is then separated into its hydrogen and oxygen constituents by electrolysis. Repeat this routine three times and you have a typical work day at *Orpheus*.

After the mining operation is finished and it is time to go home, at least 25,500 m³ of *Orpheus* will be mined and 66,000 tons of ore will be processed. The end goal is to bring back 2,550 metric tons of water as cargo. The rest of the recovered water at the asteroid will be converted to rocket propellant for the trip home. On the way back, however, the three cargo vehicles will separate from each other and go home individually. Upon arrival in low lunar orbit, the cargo vehicles will have their water unloaded, while the crew vehicle will have a changeover in personnel in preparation for the next mission. The water delivered from *Orpheus* will be converted to rocket propellant and sold on the open market. A portion of the propellant will be used to refuel the ships for the next trip to *Orpheus*.

As in the ISU Space Solar Power study, the IAM team found that this mission scenario would not be profitable in the scheme that the students devised. With a start-up cost estimated at $122 billion, the IAM team calculated that a private-public corporation would have annual costs of $5 billion with annual revenues of $5.5 billion. Nevertheless, the IAM team concluded that this type of mission is valuable and should take place. "While the mission is not viable in the present scenario, the IAM will have value in its demonstration of concept and in other benefits, like the long-term industrial and commercial advantages, economic spin-offs, scientific benefits, and establishment of infrastructure for future space missions."[121] This mission also assumes that thriving lunar surface and orbital bases will be operating. People need to realize that future space missions will never again be of the US Apollo-type in which we plant flags and make politicians and the public feel good about themselves. All future space activities are going to be integrated. People do not just go to the Moon to stand around and look pretty. A lunar base can be used for deep-space astronomy, lunar science and geology, the mining of oxygen and minerals for local use, or as a training location for Mars missions. Asteroids can provide fuel for space ships and construction material for space stations and solar power satellites. Mars can be used as a new home for human society, a jumping-off point for truly deep-space exploration, for scientific experiments, and as a potential site for terraforming. The options and possibilities are endless and our future can truly be enlightening and exciting, if we start to look toward the stars instead of at our shoelaces. None of this will happen, unless citizens begin to be involved and start speaking up. Human history seems to be in a very stagnant stage right now and few leaders and citizens seem willing to take on new challenges and risks. We continue this trend at our peril.

What About the Future?

In the last 10 years, two events have simultaneously occurred in the US space program. The first is NASA's continued operation of the space shuttle fleet. From satellite repairs and spacewalks to launch delays and the *Challenger* tragedy, NASA has succeeded in operating the world's first reusable and piloted space ship. This is what the American public usually thinks about when the space program becomes a topic of discussion. The other event, that the general public is unaware of, is the indecision the government displayed in charting the future course of our space endeavors. This can be proven by briefly looking at the large array of space reports that proposed a multitude of options for the US to pursue in space. These valuable studies have been ignored by presidents and congressional representatives. Their lack of vision for guiding this nation into the future is readily apparent. Comprehensive reports issued in 1986, 1987, 1990, and 1992 have been shelved and allowed to collect dust. This is a real shame, especially for those who are aware of the incredible promise of human exploration in space. Instead of more paperwork, it is time to make bold decisions based on the work and research that has already been accomplished.

In 1985, President Ronald Reagan formed the National Commission on Space (NCOS) and gave it the mission to chart a 50-year course for America's future in space. The distinguished group of people who compiled this report, included Neil Armstrong, former NASA Administrator (and the late) Thomas O. Paine and Chuck Yeager. They could easily claim that they produced the most comprehensive future space agenda ever published. Entitled *Pioneering the Space Frontier*, this report looked at virtually every aspect of space travel: lunar bases, Mars missions, space science, transportation systems, public involvement in the space program, and much more. The report revealed a dazzling vision with people working productively on the Moon, exploring the wonders of Mars, all connected by a cycling space ship concept that continuously traveled from Mars to Earth and back. Unfortunately, this report was issued a few months after the *Challenger* accident and became lost in the shuffle as NASA struggled to fix the problems in its space shuttle fleet. The inaction in following up on the NCOS report recommendations established a norm as subsequent space studies issued and received the same treatment. In the eyes of typical politicians, it is politically safer and easier to do nothing than to do something with a purpose and vision.

A year later, in 1987, former astronaut Dr. Sally K. Ride headed up a NASA effort to set a future course for the agency. Bearing the title Leadership and America's Future in Space, this report recommended four possible areas of focus for US civilian space efforts. Included among the options were a satellite system to study Earth's global environment, an emphasis on robotic space exploration of the solar system, a permanent lunar base, and piloted missions to Mars that would lead to permanent settlements on the Red Planet. To the authors' credit, one of their options is being pursued. The Mission to Planet Earth, which includes NASA's Earth Observing System discussed earlier, became a popular option that politicians could not resist. This effort fit in nicely with the growing concern about the health of Earth's global environment. It ultimately will teach us more about how the planet works and gauge the impact of future global warming. The other three options, to no surprise, were generally ignored.

Now is a good time to point out some of the failings of our political system. Elected national politicians usually support policies and programs that they or their constituents support. They are unable to support activities that are politically risky even though an activity might be of great benefit to the world. Our political system thrives on interest groups, lobbyists, and the entrenched interests of those already in power. Politicians also presume that the public is well informed and well educated. This is not generally the case, in this author's opinion. When an uninformed public elects uninformed people into office, the will of the ignorant is carried out more often than the will of truly informed people. That is the way it is for space. Most presidents and congressional representatives have backgrounds as lawyers and business persons and do not have an adequate understanding of future space activities and their potential. It is doubtful that this state of affairs is going to change anytime soon.

President George Bush established the Committee on the Future of the US Space Program in the summer of 1990. Led by the chief executive officer of the Martin Marietta Corporation, Norman Augustine. This group of experts formed largely as a political response to temporary problems at NASA. The year 1990 did not go down as a banner one for NASA. It began with NASA realizing that its Galileo Jupiter probe had a stuck high-gain antenna that had not completely unfolded after launch. The mission will not be a complete loss, but all the planned observations will probably not performed during the probe's orbit of the gas giant that began in December 1995. Then, in April, deployment of the much-touted Hubble Space Telescope occurred without a hitch. However, the telescope was released with a primary mirror that had an ever-so-slightly incorrect shape. This proved to be a huge embarrassment to NASA, especially after it was discovered that the mirror contractor, Perkin-Elmers Corporation, incorrectly used an instrument called a null corrector to test the primary mirror.

Adding the problems, the space shuttles Columbia and Atlantis began leaking hydrogen fuel from late May to September. This stopped any shuttles from heading for orbit. Amid these difficulties, the politicians jumped to the conclusion that another study would solve the problems at NASA. NASA fixed its problems and another study was produced to collect dust. The study concluded that the space program should be mainly devoted to science and that a heavy-lift expendable launch vehicle should be designed and built. The latter recommendation only survived a couple of years in Congress until the National Launch System was officially killed in Fiscal Year 1993. The report's first conclusion was that science should guide the space program. It is flawed in the sense that it ignores commercial space development, more affordable launch systems, and the need to open the space frontier to the common citizen. This report was of little consequence because it, too, was ignored.

The final major commission to study America's future in space, the Synthesis Report, came about as a result of the failed Space Exploration Initiative. Be forewarned that the last report reviewed here is just the same story with a different verse. The result is the same: no one takes action and America continues to sit around wondering what to do.

The Synthesis Report

President George Bush made a bold announcement on July 20, 1989 that the US would once again lead the world in space exploration. We would return people permanently to the Moon and embark on an adventure to Mars. Following this announcement, NASA did a preliminary 90-day study on the general technical and financial challenges of what became known as the Space Exploration Initiative (SEI). Once a total cost of $400 billion surfaced, Congress choked on the figure and decided SEI not worth doing. Only $5 million were appropriated to NASA's Office of Exploration before being disbanded a couple of years later. Congressional and public support for SEI never developed. President Bush hurt his cause by not including means to pay for the program. Vice President Dan Quayle then decided to conduct another study outside NASA to determine if there were better and more cost-effective ways of initiating SEI.

Thus, the Synthesis Group was created. Led by former astronaut Thomas P. Stafford, this panel of 23 senior space professionals from various federal agencies including, NASA, the US Army, US Air Force, US Marines, US Navy, the Office of the Secretary of Defense, the Department of Energy, and the Department of Health and Human Services. The specific purpose of the Synthesis Group was to take the results of the previously formed Outreach Program to determine the approaches to SEI. The Outreach Program searched for innovative ideas across the US soliciting proposals from citizens, businesses, and government on potentially improved ways to explore space.

At the Outreach Program's conclusion, the Synthesis Group designed at least two plans to accomplish SEI. When Vice President Quayle received the final report on May 3, 1991, the Synthesis Group had created four SEI scenarios. Although this report has essentially been forgotten, it is instructive to review its work to see how our future space effort may continue.

In the first program, called Mars Exploration, the goal is to explore and conduct science activities on Mars. Before going to Mars, however, equipment will need to be tested and a safe landing site will have to be chosen. The adventure begins in 1998 as two robotic probes are sent to Mars to take high resolution photos of 12 possible landing sites. A primary and backup site are then selected. A rover is dispatched to Mars in 2003 with the mission to land and travel to both sites to verify their suitability for a landing. Meanwhile, the Moon becomes the center of activity, serving as a testing site for all the vehicles and equipment to be used on Mars. From 2005-2009, the scale of lunar activities increases and leads eventually to a simulated Mars mission on the Moon. The design envisions the use of different spacecraft to carry cargo and people. The first human return to the Moon in 2005 after a cargo vehicle delivers such essentials as a habitat, power supply, food, water, and an unloader. Shortly thereafter, a crew of five lands on the Moon while a sixth astronaut remains behind in orbit. This mission, and another in 2006, will last 14 days each and will be responsible for setting up the habitat and scientific instruments at different locations on the Moon. More missions and testing take place until it is time for the rehearsal flight. The 2009 practice Mars mission begins with the crew orbiting the Moon for 120 days to somewhat simulate the Earth-to-Mars trip. The actual flight time between the two planets will be longer. The crew then lands and spends 30 days doing the sorts of activities that they will do on Mars. Once done, it is time to return home. Remembering that this scenario is science-oriented, the type of work that will take place will include exploration, digging, rock collecting, and setting up of different kinds of instruments. The real Mars mission replicates the rehearsal flight in 2012 when a cargo vehicle departs Earth. It contains a habitat, nuclear power system, and other necessities. The first Mars crew will be launched in 2014 and will arrive at their mini-camp ready for what will undoubtedly be an exciting 30-day visit. Later missions will last 600 days so that in-depth science can be conducted. Although this conceptual mission ends here, it is only logical for more people to go to Mars for longer periods of time. Eventually, the first Mars colony will be born.

The second scenario, called *Science Emphasis* for the Moon and Mars, is a variation of the first mission. Here, the Moon and Mars will be explored equally by humans and robots, the mission will be more challenging than the above mission, and a greater emphasis is placed on using the local resources available to make the mission a success. Mars operations will be a little more extensive than the first plan. For example, Mars will be the recipient of eight automated surface stations that will land at equidistant locations so that the entire network circles the planet. These stations will monitor the local Martian weather, the chemical composition of the surrounding soil, and they will provide information that will aid in deciding where to land the piloted ship. The next step is the landing of the first crew for a 30-day stay. Subsequent visits will last about 600 days each. By staying on Mars for such long periods of time, a great deal of science data will be generated. It is only appropriate now to suggest that Martian fossils may be uncovered.

A discovery this profound will shake human philosophies and religions because many human belief systems claim that people are divinely inspired and created. Many people, however, disagree with this notion. The discovery of extraterrestrial fossils will be an ultimate statement for two issues: 1) humans are not special creations, and 2) it is unlikely that we are alone in the universe. It does not matter what the fossil's size or complexity is;just knowing that it exists will be an amazing discovery. After the two long-duration visits, a permanent base is envisioned that will house 12 people. Each crew will work in 600-day shifts. Having a permanent presence on Mars absolutely requires using the Martian atmosphere and polar ice caps to produce rocket propellant and water. The Martian poles contain a significant amount of frozen water and should be very useful when people arrive. Its atmosphere is mostly carbon dioxide and this can be used to retrieve oxygen. Work is ongoing in the space community to test and perfect the technology needed to chemically alter Martian and lunar resources into useful products.

Meanwhile, the Moon is also turning into a busy place. As on Mars, surface stations will dot the lunar terrain to investigate the territory. Three piloted missions will then land at three preselected sites for 14 days to deploy telescopes and investigate the local area with the use of a lunar rover. The lunar science capabilities will expand rapidly as the next mission sets up a crew-tended habitat and nuclear power supply that will support a 90-day mission. More sensitive and capable telescopes will be constructed by the crew. Astronomy on the Moon is very attractive due to the its lack of an atmosphere. Astronomers will not have to deal with an atmosphere that causes observations to be blurry and not as distinct as otherwise possible. They will have a clearer view of the planets, stars and celestial phenomena. Also, another lunar advantage is its stability. Seismic detectors placed on the Moon by the Apollo astronauts have found Earth's satellite to be a very quiet place. The only seismic disturbances detected seem to have come from affecting meteorites.

There are many kinds of telescopes, but the crew on this hypothetical mission will build and deploy a transit telescope and a very low frequency (VLF) array of telescopes 10 km and 30 km away from the base, respectively. Each successive mission has the goal of expanding human presence on the Moon while also increasing the amount of scientific work performed. Amidst all this, a practice Mars mission is scheduled. As lunar missions get longer and longer, it will become necessary to use the local lunar resources to support base operations. Oxygen, minerals, and construction material derived from the lunar dirt will do much to make the base at least partially self-sufficient. As a footnote to this second mission idea, the Synthesis Group included an option of exploring a near-Earth asteroid with robots and humans. This is a wise consideration.

In the third report scenario, the primary goal is to construct and operate a permanent lunar base. Some work on Mars does take place, and happens according to the Mars Exploration plan, but the definite tilt is toward the Moon. Much of the early lunar activity is similar to that already discussed, but the proposed base will eventually be self-sufficient. That will be no easy undertaking. They will have to process nitrogen and oxygen from lunar dirt to create an acceptable base air supply. The base will grow or raise all the food for the crew. The base will have plants of all types to produce oxygen and to control carbon dioxide levels. Food crops may be grown in a greenhouse at the base.

Eighteen people will call the Moon home for a year. Therefore, the base should be somewhat comfortable and designed in such a way that the crew enjoys its stay. Waste management is also another vital component of a self-sufficient base. Human and animal waste will have to be recycled or used to the greatest extent possible. Human urine can be processed into clean water, although telling someone where their glass of water came from might make them squeamish. Feces could be used as fertilizer in the greenhouse. Such closed-loop life support work will be needed. Expenditures can thereby be lowered, reducing the number of resupply missions from Earth. Apart from the three habitat modules, a network of roads and landing pads will be created to increase access to the base and facilitate local travel. In time, the lunar base could eventually grow to become a small colony. Mars will likely predominate as the preferred home of future settlers, but some hardy individuals will probably choose the Moon for their own personal reasons.

The Synthesis Group's final scheme is space resource utilization. As in the other options, both the Moon and Mars will receive contingents of human explorers. It is arguable that this last plan is the most ambitious effort proposed. It also offers the greatest potential return for humans. Ultimately, the Moon and Mars will be permanently inhabited by people. The Synthesis Group recognized that the best way to succeed, is to use as many of the available resources on the Moon and Mars as possible in support of the bases. Any intelligent future space plans will include extraterrestrial resource use, to cut costs and improve space-based capabilities. This plan will pursue science, but it emphasizes resource use. On the Moon, one of the first robotic missions will place an automated resource processing plant on the surface. This machine will be tested to see how feasible it is to process oxygen and minerals from the lunar soil. When the first crew of six astronauts arrives at the Moon, their goal will be to see how easy it is to refuel an unpressurized rover and lunar ascent/descent spacecraft with the propellants processed from the lunar soil. Key requirements for the plant include reliability, ruggedness, ease of use, and ability to produce the required amount of propellant. Once these criteria have been met, the next mission phase can begin. The permanent base will support 12 people, but most of their work will be devoted to the Moon-based construction of a variety of items. They will produce silicon solar cells and iron/titanium beams on the Moon. These materials will be used in a power beaming test of microwaves that takes place between an orbiting ship and Earth. Successful transmission of electricity from the Moon to Earth may lead to larger tests and eventually a high-output system. In addition, Helium-3 will be mined in small amounts and sent to Earth. Helium-3 is an isotope of the more common Helium-2. The isotope is very rare on Earth, but it is more abundant on the Moon because of the solar radiation that is constantly bombarding its surface. Fusion research on Earth is in its very initial stages, but Helium-3 can be tested to determine it utility in a fusion reactor. As operations gear up, both rovers and space ships will be refueled on the Moon. On Mars, different crews are carrying out the same type of work. One of the first cargo flights will deliver a habitat, a pressurized rover, and an atmospheric reduction plant. This plant uses the Martian air, plus hydrogen, to make oxygen and methane that can be liquefied and used as fuel. Thirty-day and 600-day astronaut stays will follow. Later Mars missions will bring an expansion unit for the processing plant and a greenhouse. Looking down the road, some adventurous people might begin the first round of planetary migration from Earth to a new home on Mars.

Chapter 5: Space Travel and You

This chapter deals with space transportation, undoubtedly one of the most pressing issues. The very nature and capability of our transportation means dictate activities and abilities in space in the future.

The means for travelling into space should also be important for the average citizen. Until now, space exploration has been a big government and big business activity. Space missions have usually been dictated by the needs of US foreign policy, national prestige, scientific research and military security objectives. However, we are at the dawn of an era in which the space community has realized that truly expansive activity in space can only take place when space travel is accessible to everyone. The day has to come when a trip to earth orbit or the Moon is just as safe and inexpensive as air travel is today. This goal that will eventually make space travel an accepted part of human society. If the status quo continues and access to space is restricted to government and business, the space community risks permanently alienating the general public. The public will then call for a reduction or possible elimination of space activities. This is why space transportation issues and trends are so important. With the emergent goals in NASA and industry of working to innovate and reduce launch costs, the day will come in the not-too-distant-future that a weekend getaway to an orbital hotel could become a reality.

Ask any average citizen about the space program, and the space shuttle is probably the first thing they associate with it. The space shuttle era has been with us since 1981 -- and pre-flight development goes back to the 1960s. This chapter examines the space shuttle in a slightly different light. First, what were the original goals of the shuttle? Were those goals realistic? Were they met? Public policy at the federal level is rarely as clear as a pure mountain stream. Unfortunately, the aims of some public programs are as murky as the muddy Mississippi River. If the shuttle has met its original goals, what has the program accomplished, and where will we go from here? The shuttle has dictated what we can do in space and has enabled some activities, while limiting others.

The second part of this book will review some of the recent and exciting activities in the space community, such as attempts to design and construct space ships with a radical new approach. It is based on dramatically lowering the cost of space flight for people and payloads. Although real work in this new paradigm is just beginning, it bodes well for the dawning of a real Space Age in which space travel is not just the domain of a few governmental civil servants. The solar system will swing its doors open for all human society, if this enlightened policy approach is allowed to succeed.

Space Policy Goals

It seems that our national leaders have not developed or thought about the fundamental purposes or aims of space endeavors. The lack of a clear national space policy dating back to the Apollo era is one good indicator of this. There are probably many reasons for this predicament: NASA wants to do one thing, space scientists want to do another, politicians want something different altogether, and the general public is viewed as a group of cheerleaders standing on the sidelines. Instead of ignoring the role of the public in the space program, they should be the dedicated purpose of all our space efforts. US space policy -- as stated at the presidential level -- should unequivocally pronounce that its goal is *to drastically increase access to space for citizens, businesses, and the government in as short a time period as possible.* This clear goal is predicated on two simple reasons to leave Earth: to *explore the unknown and to improve the life of the peoples of the Earth.* These two items need to be the guiding lights for our extraterrestrial exploits.

How will increasing access to space for everyone help promote the two goals above? By allowing more people to travel in space, and ultimately to Mars and beyond, we will able to explore it at a much faster rate than present. It will allow commerce and industry to do work in space to provide services to Earth (solar power, etc.) that will hopefully benefit all the people of the planet. It will continue the great human legacy of expansion and settlement. The solar system is our backyard and it is just waiting for us to step outside from the back door of our global home.

These are the concepts and the goals. A reality check is now in order -- this is definitely not going to happen overnight. Thanks to the disarray in the space scene, it is unlikely that the above goal will be realized for another 10-20 years. However, we are seeing movement in the right direction.

The Space Shuttle

An entire volume could be written on the history of the space shuttle and how we ended up with the vehicle that we did. However, this book examines the shuttle only since flights began in 1981. To judge the success, failure, or maybe some fuzzy middle position of the shuttle program, we need to set the program goals. What were they? Some of the main goals were to 1) operate a reusable launch vehicle, 2) promote national security, 3) reduce launch costs, and 4) increase the flight rate. NASA may or may not have explicitly stated these goals in the early years and, furthermore, goals have changed as the shuttle era progressed. After the *Challenger* accident, the goals of conducting science and building the space station came to the forefront, while the original goals largely fell by the wayside. What follows are some of the accomplishments of the shuttle era and an analysis of whether any of the original goals were attained.

Reusability

The creation of the space shuttle was a drastic departure from earlier methods of operation. The *Mercury, Gemini, Apollo,* and *Skylab* missions all involved the use of expendable launch vehicles (ELV) to deliver people into space. ELVs are just modified missiles whose original purpose was to send nuclear warheads to the former Soviet Union. The essence of ELVs is non-reusability. Once each stage of an ELV burns out, it either drops off into the ocean or becomes a piece of space junk in orbit. Every mission requires a new rocket. With the arrival of the space shuttle, non-reusability gave way to what can be considered partial reusability.

Although some original shuttle designs promoted full reusability, the product only partially fulfilled that goal. The orbiter, which houses the crew and the flight payload, returns and lands after each mission and is one of the reusable portions of the system. The other reusable part consists of the two solid rocket boosters (SRB) that provide primary vehicle thrust in the first two minutes of flight. Both the orbiter and SRBs are refurbished and reused. The external tank (ET), on the other hand, is the one expendable part of the system. After exhausting most of its supply of liquid hydrogen and liquid oxygen that feed the shuttle's main engines during launch, the ET is released and allowed to burn up and usually lands in the Indian Ocean.

Although not fully reusable, the shuttle system has served the essential purpose of shifting the transportation paradigm from non-reusability to reusability. ELVs will never be economical enough to allow the rather increased and vast access to space that the original goal envisioned. It does not make economic or common sense to destroy the means of transportation after each use. In this sense, the shuttle has accomplished something very important. By changing the views of how spacecraft should be built and operated, the shuttle program has laid the foundation for later spacecraft that can be fully reusable as well as economical to operate.

Access to Space

Access to space is ultimately the most important aspect of any space program. If one cannot get there, then there is a real problem. If one can only get there under certain conditions, it is not much better. So it goes with the space shuttle. If we consider the number of shuttle flights and the number of people taken into space compared with previous programs, then the progress looks pretty impressive. If one considers shuttle access to space to the goal of opening the space frontier to all people, the shuttle falls way short.

Through 1993, 59 shuttle flights have carried 238 people into space.[122] Adding up the flight time from NASA mission summaries of shuttle flights, shuttles have logged about 411 days in space through 1993. By comparison, this is a dramatic improvement over the days of "man in a can" in orbit. The Mercury program placed six individuals into space, the Gemini program placed 16 persons into orbit, and *Apollo* lifted 29 men to locales ranging from earth orbit to the lunar surface. This summary does not include the reflights of different astronauts. Also, Skylab placed nine individuals into earth orbit and Apollo-Soyuz saw three US astronauts dock with a Soviet *Soyuz* capsule in 1975.

From an evolutionary perspective, the shuttle system is a vital link in the transition to widespread access to space. It is also rather unrealistic to expect the shuttle to operate under the overly optimistic cost and schedule goals propagated by NASA in the 1970s. Many areas of technology for space vehicles, e.g., computers, structures, and rocket engines, were not mature or advanced enough to make any fully reusable vehicle that was also economical to operate. At any rate, the hands of Congress, NASA, and the US Air Force all shaped the capabilities and limitations of the shuttle system.

Shuttle Science

Following the tragedy of the *Challenger* accident in January 1986, the space shuttle program underwent a major revamping. Almost three years went by with many parts of the shuttle system repaired and improved before flights resumed in September 1988. Along with the mechanical work, space policy changed too. Commercial satellites -- a major justification for the shuttle -- were banned. From then on, the major activity on the shuttle has been scientific research. Along with the deployment of a variety of science satellites and probes, much shuttle work has been dedicated to life sciences (learning how humans and other creatures adapt to the space environment) and environmental monitoring of our home planet.

From 1981-1993, shuttles have successfully deployed seven major science payloads -- six since flights resumed in 1988. These include the *Earth Radiation Budget* satellite, the *Magellan Venus* probe, the *Galileo Jupiter* probe, the Hubble Space Telescope, the *Ulysses* probe to the sun, the Gamma Ray Observatory, and the *Upper Atmosphere Research* satellite. The *Magellan* probe has completed its primary mission and has mapped most of the surface of Venus with the aid of it synthetic aperture radar. *Galileo* rendezvoused with Jupiter in December 1995 and is now circling that planet. *Ulysses'* mission is to study the sun's poles -- something never previously attempted. Most probes fly along the plane of the ecliptic until they arrive at their destination. *Ulysses* is currently traveling in uncharted space as it studies the poles of the sun. NASA's Compton Gamma Ray Observatory has the objective of peering out into the universe to search for gamma rays and their sources. Gamma radiation has very short wavelengths and is invisible to the human eye. Everyone has heard of the Hubble Space Telescope and its adventures, but we will save that story for a little later. Finally, the *Upper Atmosphere Research* satellite has been orbiting Earth and studying the different parts of the atmosphere.

In particular, however, several shuttle missions have been dedicated to studying particular scientific fields. Two of the more active areas of inquiry include the life sciences and the environmental sciences. Life science work is extremely important because of the effects of the microgravity environment on the human body. With bones getting brittle, body fluids shifting, and occasional nausea, it is important to determine the causes of these symptoms. Remedies are needed to allow people to stay in space for long durations in good health. Many shuttle flights have carried out considerable medical testing, but two missions were exclusively devoted to poking, prodding, and generally examining the human body in space.

These shuttle flights, called the Spacelab Life Sciences (SLS) missions, were dedicated to answering many of the fundamental questions about the human body's reactions in space. Starting with the STS-40 mission of the shuttle *Columbia* in June of 1991, astronauts became the dedicated guinea pigs as coordinated, space-based life sciences research began. Work on SLS-1 began in 1978 when NASA started its search for principal investigators (PI) for the mission. PIs are scientists on the ground, from both academia and industry, whose experiments have been approved for flight on the shuttle. Whether it is astronomy, earth sciences, life sciences, or other fields of study, the PIs are the people in charge of the experiments flown on the shuttle or on any other NASA scientific mission.

During this nine-day mission, 18 experiments were set up that involved both humans and animals. Along with the crew of seven, there were 30 rodents and 2,400 jellyfish who came along for the ride. The following NASA press release addresses the SLS-1 mission.

An example of the SLS-1 experiments will indicate the nature of this mission and some of its tentative results. It is very important to realize that a single experiment does not provide any definitive results. Experiments need to be performed, repeated, and published in the scientific community for appropriate scrutiny. The scientific method is a continuing process and initial experiments at best provide tentative evidence for or against a given hypothesis.

The five experiments on SLS-1 that we will review were designed by Dwain L. Eckberg, Leon E. Farhi, John D. West, Clarence Alfrey, and Claude D. Arnaud. All are medical doctors at colleges and universities across the United States. Some experiments have different goals, while others may seem quite similar. For example, Eckberg wanted to understand a theory that claims that the normal human reflex system for regulating our blood pressure works differently in space. This was a rather simple experiment in which astronauts had their blood pressure checked before, during, and after their mission. While in space, astronauts wore a neck device that measured the blood pressure in their neck.

Farhi, who works at the State University of New York (SUNY) at Buffalo, designed an experiment for cardiovascular deconditioning, a process in which the heart weakens in space. Without the normal unity-gravity force we feel on our body, a person's heart in space does not have to work as hard. This is an issue for long-duration space missions because this weakened state of the space traveler could become serious on reentering a strong gravity field. Cosmonauts from the long-duration Russian missions on the *Mir* space station have to be carried out on stretchers after their mission because their bodies are exhausted by their extended stay in space. Russian missions have lasted for about a year or less; longer missions will require astronauts who can counteract this problem.

In particular, Farhi wanted to measure how quickly the human body adapts to microgravity and then how quickly it re-adapts to Earth's gravity. To perform this test, astronauts were instructed to re-breathe the air they had previously exhaled. From this measurements could be taken on the amount of blood pumped out of the heart, oxygen usage, the amount of carbon dioxide released by the body, heart contractions, blood pressure, and lung functioning. This re-breathing activity took place in the orbiter's mid-deck as astronauts rode an exercise bike. Measurements were also taken with the astronauts at rest.

Dr. West's experiment aimed to determine how human lungs operate in microgravity. Is the flow of oxygen into the bloodstream the same, higher, or lower while in space? What is the rate of blood flow into the lungs while in space? These are the types of questions this experiment was attempting to answer. To do this, the chosen astronauts were required to conduct eight different breathing tests, using bags that contained specific gas mixtures. By breathing into these bags, it was possible to see if lung capacity changed after exposure in space.

Meanwhile, Dr. Alfrey, a physician at the Baylor College of Medicine at Houston, wanted to establish the causes for the lack of red blood cell production in space. He also wanted to measure the blood 's oxygen carrying capacity. Blood samples were taken before, after, and during the flight to measure and compare red blood cell production. Chemical tracers were injected into certain astronauts before launch to measure the destruction rate of the red blood cells. These chemicals, which were injected on the mission's second day, attach themselves to individual red blood cells and make it possible to track their life and death. Blood samples were taken throughout the mission to chart the process of red blood cell production and destruction. The results of this experiment proved interesting. Red blood cell production in orbit did decrease. However, the blood cell life span remained the same and hemolysis -- the disintegration of red blood cells – was absent. Also, red blood cell mass and plasma volume experienced decreases in orbit.

A final SLS-1 experiment was Dr. Arnaud's work designed to measure circulating levels of calcium metabolizing hormones and the inflow and outflow of calcium from the body. Even on Earth, calcium is an important part of a person's diet. Osteoporosis is a condition in which bones become porous, brittle and prone to fracture. It affects some 25 million women, specially elderly women. In space, the problem is even worse. For some reason, extended space flight reduces the people's ability to keep the calcium in their bones. This continued bone demineralization could be a big problem for future space travelers who will have to spend months in space going to Mars and beyond using conventional propulsion methods. A remedy is needed to assure the health of astronauts in space, particularly if people are ever going to migrate to other moons or planets in large numbers. This experiment involved daily weight measurements of all seven of the astronauts and records of all the food, liquid, and medication that each person took in. Blood samples were also taken from each crew member and the same experiment was conducted on a control group of subjects on Earth. The results of this test showed one of the negative health effects of having people in space. It seems that the bone-dissolving cells in the body worked at a higher rate than the bone-forming cells in space. In addition, the ionized calcium blood levels found in space were considered clinically abnormal for a person on Earth.

This is just a sampling of the work on SLS-1 and some of the preliminary results. Each experiment has played a small part in understanding more fully how our bodies work in space. Solutions to some of the problems encountered will have to be developed to allow healthy long-term space flight. Alternatively, effective methods of artificial gravity have to be applied to future deep space vehicles. Below is the NASA press release mentioned earlier about the results of one of the animal experiments on SLS-1.

Shuttle Rat Experiment Yields Unexpected Results[123]

Unexpected results from a recent Space Shuttle experiment on rats someday may lead to the discovery of genes that direct bone cells to produce more bone. Last April's third Physiological and Anatomical Rodent Experiment (PARE-3) [in 1991] studied changes in the activity of bone-forming cells after 9 days of space flight. It also investigated whether these changes were reversed within 3 days of return to Earth.

"We were totally surprised by some of the data from this flight experiment", recalled Dr. Emily Morey-Holton of NASA's Ames Research Center, Mountain View, California.

Although she cautioned that much research remains to be done, she thinks the findings may accelerate the development of drugs to stimulate the production of new bone. Such medications could be important not only for astronauts on long-duration space missions, but also for people on Earth suffering from bone-weakening disorders.

At the end of the mission Morey-Holton and co-investigators Dr. Kim Westerlind and Dr. Russell Turner of the Mayo Clinic, Rochester, Minnesota, found a decrease in the production of certain chemical messages necessary for bone formation. This was followed within 24 hours by a dramatic increase in production of these chemicals. Levels again decreased within 72 hours after landing.

The 40 percent decrease in chemical messages was seen in cells on the outer surface of the lower leg bones (cortical bone). These messages direct bone-forming cells to produce proteins that are part of the process that results in formation of new bone. The investigators had expected this decrease since space flight is known to slow the rate of bone growth.

"The big surprise was an unexpected 300 percent increase in these proteins within 24 hours after return to Earth", according to Morey-Holton. "The response to reloading must have occurred almost immediately." Walking or movement against the force of gravity is known as a load.

Morey-Holton postulates that re-exposure to Earth's gravity may have caused the cells to overcompensate for the slowed bone growth during space flight. "This suggests that the best time to learn about the initial steps in bone formation may be within 24 hours after space flight. The increase is probably large enough to allow scientists to establish a time course of events during the 24-hours period", she estimated.

"It may be possible to tease out the genes involved in initiating cortical bone formation. If you know what the actual trigger is, you should be able to use that particular message to stimulate bone formation", she noted. This, in turn, could lead to the development of drug therapy.

Because rats with unloaded hind limbs on Earth "showed virtually identical responses to the flight animals" on PARE-3, much additional research can be done in laboratories on the ground, she explained. "This should accelerate the research, because it eliminates the years of waiting necessary for a flight opportunity on the Space Shuttle."

SLS-2

A year later the second Spacelab Life Sciences mission went aboard a shuttle, again using the orbiter *Columbia*. This adventure, designated STS-58 also known as the mission with the headless rats, became one of the longest in shuttle history.

As on SLS-1, a contingent of 48 rats were lifted into orbit and underwent some of the same experiments as the first mission. The only difference was that six of the little critters had their heads chopped off by a miniature guillotine, so they could be dissected immediately following death.

Some of the SLS-1 experiments were repeated -- a necessity in any scientific endeavor -- and a few were new. There were 14 experiments on SLS-2, eight involving humans and six involving rats. To continue the review of the life sciences I will describe two experiments on SLS-2.

Dr. C. Gunnar Blomqvist of the University of Texas Southwestern Medical Center in Dallas proposed an experiment similar to Dr. Farhi's investigation on SLS-1. Instead of focusing on the entire human body, Blomqvist was interested in the functioning of the human heart. He asked how the human heart changed during an operation without gravity. This experiment also flew on SLS-1 and the results proved intriguing. Central venous blood pressure showed a decrease in astronauts in space. This was a surprise because a blood pressure increase was predicted. This increase was expected because of the shifting of body fluids that takes place in microgravity. These experimental results invalidated the theory. Also, astronauts had an increase in their heart size, more blood was pumped per beat, and the subsequent blood flow also increased. With more experimentation needed to confirm these findings, the investigations will eventually determine if any of these conditions are harmful in the long term..

Finally, T. Peter Stein performed an experiment to determine why astronauts lost muscle mass in space. Intuitively, it makes sense to expect that the lack of gravity reduces the need for strong muscles in space. Even normal activities in space do not use many muscles and this lack of use results in atrophy. It would undesirable to journey to Mars for eight months, land there, and then to find it difficult to walk on the surface. Stein specifically looked at the changes in the body's protein metabolism that leads to muscle loss in space. SLS-1 results bore out the above intuitive notions. It seems that protein synthesis and breakdown increased in space, but the protein breakdown took place at a higher rate. On SLS-2, further studies were conducted to verify the earlier results. Some crew members were supposed to swallow an amino acid that contained a non-radioactive isotope of nitrogen. This isotope enabled one to trace the protein metabolism process in the astronaut's body. After swallowing the amino acid and waiting about 10 hours, the crew took samples of blood, saliva, and urine for later analysis.

It is to be hoped that the description of the life sciences experiments has provided a representation of the investigations and demonstrated their importance. Any long-term human migration in space will need a thorough understanding of human biology in space. What is more important, remedies to the adverse medical conditions in space will have to be invented or artificial gravity will have to be designed into future ships. This is not frivolous work -- it is vital to our future in space.

Shuttle Rescue and Repair Missions

Most of the shuttle missions currently flown involve objectives such as those discussed above in the life sciences section. Regardless of the scientific goals, these types of missions generally tend to evoke no interest from the public. However, when a shuttle mission is scheduled to rescue or repair a stranded satellite in orbit, the public reacts with intense interest and curiosity. Unlike science missions, most people can understand the objective when a team of astronauts is assigned to repair a malfunctioning satellite. Our planes, trains, and automobiles all break down occasionally and many people know their mechanic by his or her first name. In the case of the shuttle, the repair team is a group of astronauts who are zipping along hundreds of miles above us. It is a different place, but the same mission.

Many more millions of people followed a series of very high-profile shuttle rescue missions than any normal mission. The press also described them as possibly crucial to NASA's future. The most vivid examples are the most recent missions. During the repairs of the Intelsat 6 satellite on STS-49 and the Hubble Space Telescope on STS-61, space experts were quoted that failure to accomplish the mission could damage the future of the space program.[124] These messages were very strange. It is hard to imagine why the success or failure of one mission could ultimately determine the future of our space efforts. The reason for these attitudes and public statements may reflect the problems that NASA has faced since the *Challenger* accident. Some major missions have failed or been partially crippled (*Hubble, Galileo, Mars Observer,* etc.) and NASA is sometimes seen as a typical bureaucracy that wants to protect what it has (the shuttle, space station, etc.) without boldly going where it should be going (Mars, the Moon, economical launch vehicles). One must remember, though, that NASA ultimately receives its guidance from the White House and Congress. Without proper guidance from our elected leaders, it is hardly reasonable to blame all NASA's problems on NASA. Realistically, the blame should be spread throughout our government to include past and current presidents, Congress, and NASA.

The type of activities during satellite rescues are exactly the kind we need to pursue in space to increase our capability to conduct larger operations. It is vital for us to be able to work and build structures and habitats beyond our home planet, if we ever plan to explore and visit the rest of the universe on a permanent basis. In a few years, NASA and Russia will begin constructing an international space station. This will be a huge project and will test our international ability to work in space effectively. Solar power satellites are magnitudes more difficult in complexity and size. Lunar and Mars bases also will be difficult missions with our present abilities. The bottom line is that our current capabilities in space are not directed at helping the human species expand into the solar system. This must change. A look at the shuttle rescue missions will indicate the kind of work we need to conduct in space. It will point out the requisite capabilities to begin our migration from our home planet. Four missions will be summarized: STS-41C, STS-51A, STS-49, and STS-61. These missions involve the retrieval or repair of spacecraft in orbit that have malfunctioned in one way or another. It is important to remember that any future piloted space ships should be capable of these types of missions -- and more. Much work can be performed in space, but we need to be able to do any productive work.

STS-41C

The first shuttle rescue mission (STS-41C) took place in April 1984 when the shuttle *Discovery* and a crew of five were assigned to repair the *Solar Maximum Mission* (SMM) satellite. Launched onboard a *Delta* rocket from Cape Canaveral on February 14, 1980, the SMM's mission was to study the periodic solar flares. A unique vehicle in its own right, the SMM was of the first satellites designed to be retrieved and repaired by a shuttle crew. Problems arose on the satellite at the end of 1980 and into 1981, when three sealed fuses in SMM's attitude control system failed. This control system was important because it helped point the satellite's scientific instruments precisely at various parts of the sun. Without a healthy attitude control system, ground controllers subsequently put the satellite into a sun-pointing mode and gave the vehicle a slow spin. The problem with this maneuver was that it made four of the instruments onboard unusable because they could not be focused on their intended target. SMM therefore became the target of the first shuttle mission scheduled to fix an orbiting satellite.

Led by veteran astronaut Robert Crippen, *Discovery's* crew also included Dick Scobee, George "Pinky" Nelson, Terry Hart, and James van Hoften. After launching into orbit and deploying the Long Duration Exposure Facility (LDEF), the crew began the process of rendezvousing with the wayward spacecraft. Three major repair events were scheduled: replacement of the attitude control module with a spare unit, replacement of the main electronics box for the Coronograph/Polarimeter instrument, and the installation of a cover over the gas vent of the X-Ray Polychrometer.

Arriving within 200 feet of the satellite about two days into the mission, the EVA team of Nelson and van Hoften geared up to head outside to bring *Solar Max* into the payload bay. The original plan had Nelson donning a Manned Maneuvering Unit (MMU) -- a fancy jet pack to allow individual travel in space -- and flying over to the satellite to attach a capture device located on the front of the MMU with a grapple pin on the satellite. The automatic attitude control function of the MMU would help to stop the spin of the satellite. As later missions would also illustrate, the best laid plans of man and machine sometimes go awry.

After slowly approaching the satellite, Nelson went straight in to lock himself to the satellite. Unfortunately, he just bounced off. He tried again, with the same result. After a few more attempts, Nelson was ordered to return to the cargo bay. The backup plan focused on using the robot arm to snare a grapple pin on *Solar Max*. Unfortunately, Nelson's docking attempts put the satellite into a slow tumble that made a robot arm capture difficult. The astronauts tried anyway, were unsuccessful, and NASA told the disappointed crew to call it a day so everyone could regroup. A few days later the *Challenger* crew successfully captured the satellite and brought it into the cargo bay for repairs. This capture came about in large part due to efforts by ground controllers at the Goddard Space Flight Center to slow the spin rates of *Solar Max*.

Once in the payload bay, van Hoften and Nelson replaced several parts of the satellite and successfully returned it to orbit. This repair proved very important because NASA had estimated the cost of a new satellite of that type to cost about $235 million. The repair cost came to about $48 million.

This mission accomplished two goals. The first is that satellite rescue and repair in orbit by a shuttle crew proved feasible. The second is that even if problems do crop up, the crew and mission control can be resourceful in finding ways to solve the problems that confront them.

STS-51A

Seven months later in November 1984, another shuttle crew was summoned to retrieve and return to Earth a pair of satellites that had become victims of malfunctioning rocket boosters. Palapa B-2 and Westar 6, both commercial satellites, had originally been deployed on STS-41B in February of the same year. Deployment of both satellites from the shuttle took place as scheduled, but gremlins were lurking in the payload assist modules (PAMs) that were designed to send the satellites to their proper location in geosynchronous orbit. The PAMs were designed and manufactured by the McDonnell Douglas Corporation. Unfortunately, both PAMs did not fire as planned and placed the satellites into useless 600-mile-high orbits.

With a crew of five, this mission (STS-51A) for *Discovery* involved several aspects. It was commanded by Frederick Hauck and included David Walker, Anna Fisher, Dale Gardner, and Joe Allen. *Discovery* maneuvered over to Palapa first and Allen and Gardner donned their space suits to prepare for the mission at hand. The spacewalkers used a "stinger" device that was inserted at the rocket motor base to allow recovery by the shuttle's robot arm. This method worked in the retrieval of both satellites, but problems developed in stowing the *Westar* satellite. The spacewalkers had to quickly improvise to move *Westar* into the cargo bay because an A-frame could be not attached to the satellite as planned. The improvisation then had Allen holding onto the satellite until the other astronauts could move to their necessary positions to help guide in the satellite. This became the first mission ever to retrieve and return satellites to Earth. Millions of dollars were saved and the two satellites were later relaunched and entered geosynchronous orbit.

STS-49

Years passed and the excitement and adventure of satellite rescues by shuttle crews was forgotten as NASA labored to recover from the *Challenger* tragedy and to fly the shuttle again. The next opportunity to demonstrate the shuttle's abilities came on March 14, 1990, after a Martin Marietta *Titan* 3 rocket launched an Intelsat satellite. To the dismay of the satellite owner and the rocket builder, the Titan 3's second stage did not separate from the satellite in low earth orbit. In a desperate attempt to salvage the mission, Intelsat separated the satellite from its attached booster stage and then circularized the satellite's orbit a few days later.

After negotiating with Intelsat to receive $91 million in payment for the repair mission, NASA designated STS-49 as the mission to rendezvous with Intelsat and install a new booster rocket on the satellite. Set to occur in May 1992, this mission also marked the first flight of the shuttle *Endeavour*. The original satellite rescue plan had the robot arm move crew member Pierre Thuot toward Intelsat with a long capture bar in hand that he was to inserted at the satellite base. The shuttle's robot arm would then grab the capture bar and bring the satellite into the cargo bay.

With Endeavour next to Intelsat and Thuot positioned at the end of the robot arm, the astronaut carefully positioned the capture bar up against the rear of Intelsat. A strong push should have triggered the capture bar to latch onto the satellite. However, this did not occur and the force of Thuot's push allowed Intelsat to start floating away. Other attempts to get the capture bar to latch proved futile as well. The problem arose from the repair simulations that took place underwater on Earth months beforehand. The water's resistance in the training pool allowed the capture bar to latch onto a simulated satellite. In space, however, no force was present to allow the connection to occur and Intelsat just floated away. Two spacewalks tried to achieve this, but things were not looking good after the second round of efforts failed.

Human ingenuity prevailed in the third attempt. The crew proposed a scenario in which three astronauts would be in the cargo bay to simultaneously grab hold of the satellite. With Thuot on the robot arm, Tom Akers mounted on a platform in the middle of the cargo bay, and Richard Hieb located inside the starboard cargo bay sill, all three astronauts waited patiently until the slowly spinning satellite approached them. At the right time, they grabbed it. The smooth and cylindrical satellite soon sported a capture bar and the spacecraft eventually made its way into the shuttle cargo bay. Work then continued to attach a fresh rocket booster to Intelsat that later allowed it to take its place

22,300 miles above the Earth. The $157 million satellite is now in its proper home in space helping people to communicate all over the world.

STS-61

Undoubtedly the most ballyhooed shuttle mission in recent years was that of shuttle *Endeavour* in December 1993. A dedicated effort was aimed at correcting the highly publicized problems of the Hubble Space Telescope (HST). Hubble is a very large space-based telescope that orbited Earth since April 1990. Touted as possibly being able to detect planets and to see far back into the history of the universe, Hubble quickly became one of the biggest embarrassments NASA ever faced. This is because NASA found that it could not focus the telescope due to an optics error called spherical aberration. This is caused by a miss-shaped telescope mirror that does prevents the precise focusing of light. The telescope's primary 94.5 inch diameter mirror was the problem. To focus properly, the mirror had to be ground down to an exactingly perfect shape. The contractor in charge of this work came up only 2 microns off specifications and it was later discovered that the mirror's shape was incorrectly tested. The result was that a major mirror flaw escaped both the contractor's and NASA's attention and was sent into space along with the rest of the telescope.

This is not to say that Hubble became instantly useless. Several of the telescope's onboard instruments were only marginally affected by the problem and computer enhancements also helped to clear up many of the images taken by Hubble. However, HST hung around NASA's neck like an albatross and seemed to continue a string of bad luck at the agency dating back to *Challenger.* Even as the uproar settled down and Hubble began its work, system malfunctions emerged on the telescope. Gyroscopes that help to point the craft failed. The solar panels that provide electricity to the telescope wobbled violently as the telescope moved from darkness to daylight and vice versa. The panels were not designed to handle the strain of such rapid temperature changes. Some of the memory systems in Hubble's computer also failed.

With all these developments, NASA designated STS-61 as the Hubble Space Telescope repair mission. Leading up to the mission, NASA reviewed and reviewed its plans to make sure everything would go right. The press and the public openly wondered what a mission failure would do to the space program. Despite the concerns and doubts, there were some advantages to working on Hubble compared to other satellites.

The most important advantage is that Hubble was designed to be specifically serviced by the shuttle. Hubble had modular components that could be easily removed and replaced. Hand-holds were plentiful and maneuvering about was designed to be easy for astronauts. This is in contrast to the previous Intelsat rescue that was not designed to be handled by astronauts.

Endeavour thundered through the Florida skies on December 2, 1993, and the chase for Hubble began. Rising to an altitude of 515 km (322 miles),

Endeavour's crew quickly snared the telescope and placed it in the cargo bay where four astronauts prepared to conduct electronic surgery. From December 4 to December 8, two teams of two astronauts each took turns in fixing the maligned telescope. The astronauts worked over five days to install corrective optics and the Wide Field/Planetary Camera II; replace gyroscopes, computer equipment and solar panels; and tested the overall health of the telescope.

Hubble flew freely again on December 10 and *Endeavour* landed at the Kennedy Space Center on the same day. The four spacewalkers had spent a total of 35 hours and 28 minutes on the repairs. They spent that time using more than 150 specially designed tools to help them get their job done. The only nagging doubt that remained was: would Hubble work as originally advertised?

All doubts about Hubble's capabilities were erased on January 14, 1994, as NASA triumphantly showed to the world images of the newly rehabilitated telescope. Similar pictures taken before and after the orbital operation showed without a doubt that Hubble had 20/20 vision and would be able to take our species to places never seen before.

The reason for reviewing some of the more dramatic and daring shuttle missions is to show that the shuttle is truly a national resource that should be used to its greatest potential. The above missions also illustrate the ability of people working in space. An old dogma still prevails among some people who maintain that space should be the home only to robotic probes and maybe brief human forays. This dogma ignores the promise of discovery in our larger celestial home and the requisite rewards that will come from it. Soon the shuttle will begin constructing a space station above our heads. From there, we need to keep looking outward, continue striving skyward, and learning how to do more and more.

International Cooperation

One of the most enduring aspects of America's civil space program has been its foreign policy component. American presidents, from Dwight Eisenhower to Bill Clinton, all seem to have a need to justify space missions by tying these activities with the foreign policy goals of the United States.

This foreign policy coloring of space missions has taken on many different faces. In the 1960s, President John F. Kennedy challenged America to safely send human beings to the Moon and return them to Earth before 1970. This national goal grew out of *Sputnik*, the perceived nuclear missile gap, and the superiority of Soviet technology over American technology. America's push to the Moon soon became a race with the Soviets. This race had many highly symbolic characteristics. Both nations were determined to reach the Moon first to show that their economic and political systems were superior to that of their rival.

In the 1970s, detente caused a thawing of the tepid relations between the US and the Soviets. During a particularly "warm" period in their relations, the US and the USSR decided to conduct a joint docking mission to symbolize the potential for peace. In 1975, a Soviet *Soyuz* capsule with cosmonauts Alexei Leonov and Valeriy Kubasov, and the last Apollo flight carrying Vance Brand, Tom Stafford, and Deke Slayton linked up in low earth orbit for a handshake and a few experiments. This mission accomplished very little scientifically or technologically, but served as a foreign policy triumph for both nations.

A decade later produced a different type of foreign policy impact on the space program. NASA and the European Space Agency (ESA) agreed on building the Spacelab module that is carried in the cargo bay of space shuttles. ESA built the modules that were essentially given to NASA for use. Later in the 1980s, NASA's proposed space station took on a decidedly international flavor. Originally intended to include ESA, Japanese, and Canadian participation, the most recent incarnation of the space station brought in Russia as a major player. If it flies as currently configured, the international space station will be a model for international cooperation in space.

The international exploration of space is very important for several reasons. First, the final frontier is vast and forbidding and will require the collective talents of all the nations of Earth working together. No single nation will ever be able to accomplish what a global effort would be able to do. Second, international space cooperation may go a long way towards tearing down the walls of destructive patriotism, suspicion, and hatred prevalent in our civilization's past. The social aspects of this international effort may promote true tolerance, respect, and understanding for people of different cultures, belief systems, and values. Working together is much better than working apart. Space exploration could go a long way towards healing the ancient divisions between nations.

In preparation for construction of the international space station in late 1997, NASA and RSA -- the Russian Space Agency -- have agreed on a series of missions that will have space shuttles docking at Russia's *Mir* space station. These cooperative missions are designed to familiarize the Americans and Russians with each other's way of doing business before construction of the international space station begins. Initially formulated during a June 1992 summit between US President George Bush and Russian President Boris Yeltsin, the space cooperation activities have nine shuttle dockings taking place at *Mir* in the next several years. The bi-nation cooperation has already begun. A January 1994 flight of the space shuttle *Discovery* saw the first Russian cosmonaut , Sergei Krikalev, fly on an American spacecraft. Vladimir Titov, a veteran of two space flights and a year in space, flew on shuttle *Discovery* in February 1995 as the ship made the first-ever close approach to *Mir* in preparation for a June docking by the shuttle *Atlantis*.

Space shuttle *Atlantis* successfully approached and docked with *Mir* in late June 1995. Greeted by two Russian cosmonauts and US astronaut Norman

Thagard at the station, *Atlantis* spent several days docked with the Russian facility so that joint medical experiments could be conducted. Thagard, who had rocketed to *Mir* on a *Soyuz* rocket in March, became the US space endurance record-holder by staying in space for 115 days. Previously, the third Skylab crew had held the US space endurance record of 84 days. Shannon W. Lucid became the next American to spend time on *Mir* -- this time for a five-month stay. Lucid conducted her mission on Mir from March until August of 1996. The third American to fly on *Mir* will be Jerry M. Linenger, who will also arrive at the Russian station in 1996 after Lucid's mission.

Shuttle Failings

Assessments of the space shuttle program's accomplishments and failures will always be a influenced by subjective interpretation. As stated at the beginning of this chapter, the very broad space policy goals tend to cast the shuttle in a bad light. Three main shuttle objectives -- a high flight rate, low operating costs, and a high level of maintainability -- have not been accomplished. In the 1970s, NASA presented the shuttle as a do-everything, low-cost space transport that would revolutionize what we do in space. If the shuttle had accomplished those goals, the world community would be doing a lot more in space compared to what we are doing now. Intricately tied to cost estimates for shuttle operations were the annual number of flights needed to make the system economical. From 1970 to 1971, NASA hired Mathematica, Inc. to conduct a study on the economics of the shuttle. At that time based on a completely reusable vehicle, Mathematica "concluded that the fully reusable system would only be marginally cost effective. The main determining factor would be the number of flights per year. The more flights per year, the lower the cost for each flight." NASA later abandoned a fully reusable vehicle, but in the process gave out estimates of 715 shuttle flights from 1978-1990.[126] Unfortunately, reality did not prove so kind (see Figure 5).

As illustrated in Figure 5, the flight rate accelerated greatly from 1981 through 1985. In 1985, the shuttle accomplished its most ambitious flight schedule -- nine flights. The following year produced the *Challenger* tragedy and with it a shutdown in shuttle operations until September 1988. Since 1989, the annual flight rate has varied from five to eight missions per year. Clearly, the shuttle flight rate will never accomplish its original estimates. It is also clear that the shuttle will not be the vehicle that will open up the space frontier for human society.

Cost is also a factor that has hounded the shuttle system from the beginning. Federal expenditures that have gone into the shuttle system through 1992 are given in Table 4. When dealing with the cost of a single shuttle mission, however, the estimates range from cheap to quite expensive -- depending on who is providing the figures. The range of costs is mainly attributable to what

The Case for Space

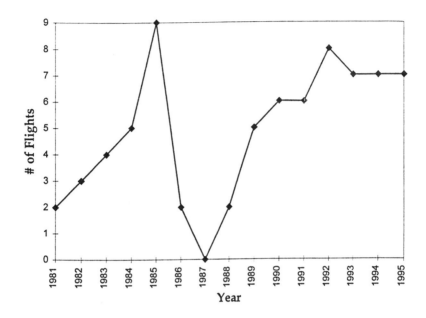

Figure 5. Annual shuttle flight rate, 1981-1995.

is included in the mission cost estimate. According to David Bates, deputy director of resources management at NASA, the marginal cost of launching a shuttle is $37 million. This includes the cost of labor, fuel, processing, and other items directly related to a specific mission.[127] NASA public affairs derives the cost of a shuttle flight by dividing the shuttle operations budget by the number of flights for a given year. For 1993, the cost using this method became $368 million per flight.[128] However, another NASA official said that a cost of $547 million per launch was accurate.[129] Outside the space agency, there are people who think the cost of a single mission is even higher. John Pike, a space policy analyst for the Federation of American Scientists, cited a cost of about $750 million per flight in 1992.[130]

Amidst all these estimates, NASA has diligently tried to reduce the cost of operating its shuttle fleet. Initially, NASA had a goal of reducing the shuttle budget by 15 percent from 1992-1996. However, as time went on NASA managers became more aggressive in their desire to reduce cost. "The shuttle management team had pledged to reduce operating costs 25 percent by 1995 -- a steep reduction of the space agency's ongoing plan to cut costs 3 percent a

TABLE 4

Space Shuttle System Expenditures, by NASA in constant (1982) dollars:
1973 to 1992 (In millions of dollars)

Year	Total	Construction	Production	Operations
1973	58	58	0	0
1974	325	109	0	0
1975	1,543	136	0	0
1976	2,619	79	0	0
1977	2,258	60	109	0
1978	2,062	103	61	0
1979	2,251	42	496	0
1980	2,751	42	907	463
1981	2,724	13	1,093	578
1982	2,932	20	1,283	735
1983	3,014	26	1,635	1,354
1984	3,127	73	690	2,364
1985	2,636	42	1,376	1,218
1986	2,606	0	1,184	1,422
1987	4,165	0	2,611	1,553
1988	2,538	0	948	1,590
1989	3,084	0	943	2,141
1990	2,996	0	968	2,028
1991	3,058	0	989	2,069
1992	3,140	0	976	2,164
Totals	49,887	803	16,269	19,679

Source: US Congress, Congressional Budget Office, unpublished data.
Statistical Abstract of the US, 1993, Table No. 999, page 606.

year."[131] Budget figures show that shuttle operations expenditures have decreased from $2.857 billion in FY 1993 to $2.420 billion in FY 1995.[132] NASA Administrator Daniel Goldin has also stated that he will decide the fate of the shuttle system by 1997.[133] Depending on Goldin's decision, the shuttle system could continue operating well into the 21st century or could be given an early retirement to make way for new, more economical piloted and unpiloted spacecraft.

It is very clear that NASA has never been able to make the shuttle cost-effective. When carrying commercial satellites to orbit in the early 1980s, NASA offered subsidized prices to attract commercial customers to the shuttle. With its main commercial competitor being ESA's *Ariane* rocket, NASA could not offer prices designed to recoup the original costs of developing and building the space shuttle. The space agency would not get any customers at all if it did that. However, it should also be noted that operating the shuttle is not a business -- it is a government activity. The shuttle was designed to meet NASA and US Air Force needs. NASA tried to, nevertheless, create an illusion in the budget-shrinking era of the 1970s that the shuttle could be economical. Its driving goal was not to reduce the cost of access to space or to open up the space frontier substantially beyond the civil servants that fly on the vehicle.

Finally, the maintainability and operability of any vehicle is another measure of its usefulness. If your car kept breaking down and it had to spend much time in the shop, you probably would consider trading it in for a better model. The analogy fits the space shuttle, but only somewhat. A car is a far cry from a first-generation, reusable space ship. The shuttle's complexity is magnitudes greater. Space technology is not nearly as mature as automobile technology, and moving to and from space is much more difficult than driving along Interstate 80. In the context of reliable transport to orbit, though, the shuttle has had its fair share of growing pains. Without discussing the numerous launch delays that show the system is incapable of meeting a schedule, there have also been other vehicle problems that have put a major crimp into the flight rate of the shuttle.

Most people are familiar with the O-ring exhaust leak in a solid rocket booster that led to the demise of *Challenger* and its crew. This technical problem, along with the management failures that accompanied it, shut down the program for almost three years. In 1990, a set of hydrogen fuel leaks grounded the fleet for about five months as NASA had to stop everything to search for the problem. During this period, hydrogen leaks appeared on the shuttles *Atlantis* and *Columbia*. The fleet was officially grounded on June 29, 1990, after a special tanking test on *Atlantis* produced hydrogen leaks from an unknown location. Initial analysis focused on the 17-inch quick disconnect valve on the underside of each orbiter that regulates hydrogen fuel flow and closes when the shuttle is ready to jettison the external tank. In July, NASA thought that the two leaks were different,[134] but later tests showed that leaks on both orbiters were similar.[135]

With all this said, what conclusions can we draw about the US shuttle system? It is fair to look on the shuttle and its abilities -- however imperfectly planned and devised -- as a bridge. The shuttle allows human society to cross over and become regular travelers beyond Earth. Maybe it can be likened to a Model T car or an old bi-plane. Its abilities to do the job are limited -- limited by its technology, its design, and its purpose. However capable the shuttle is, there is the potential to make space ships that are substantially more proficient. The shuttle is, indeed, a forerunner to a class of space ships in the future that will undoubtedly amaze us all. Maybe it has allowed us to realize that we can live and work in space; that we can continue to explore the unknown; that the surprises and unknowns of space are not unattainable. With all its limitations, the space shuttle has allowed us to extend our reach -- a reach into the universe that has hopefully just begun.

The Next Era

The singularly simple space transportation goal -- access to space for individuals, businesses, and governments -- is one of the most pressing and overriding needs of our space exploits. However, the US federal government has usually viewed space policy for national interests at the macro level. Whether for national security, economic competitiveness, or basic research, space activities are geared to benefit the general goals of the United States as a nation.

There is nothing wrong with structuring US space activities to serve the national interests of America. However, the space goals of the United States need to become more expansive and inclusive. As stated earlier, US policy makers need to look to serve the interests of all Americans. How would cheap and quick access to space benefit anyone? Space-related businesses would find their launch costs drop dramatically and they would be able to do more work in space. A dramatic drop in the cost of getting into space could spur more space-minded companies to work to get their products into orbit. In addition, the US government could do more useful work. Missions back to the Moon and expeditions to Mars could finally become feasible. More people could have access to orbit and more manned space facilities could be built. The possibilities are endless when everyone has access to it. Before seagoing vessels were built, humans were largely limited to their immediate geographical area. With the arrival of airplanes and ocean-crossing ships, people traveled all over the world.

The ability of individuals to travel to orbit, with the ease now seen in airplane travel, would familiarize more people with the space environment and its potential. At an appropriate time, the US government should maybe institute a new Homesteading Act. This Act, originally enacted in 1862 to encourage settlers in America to move west and claim land, helped early Americans realize their dreams of becoming farmers. Maybe a Space Homesteading Act could be drawn up at the appropriate time to encourage people to settle the Moon or Mars. Critics may argue that there is no need to pursue individual access to space because there is no market to satisfy. This is narrow-minded thinking. If access to space becomes as open as jet travel, then motivated people will pursue their interests and move out into the solar system.

To reach these possible scenarios, the US government needs to abolish its policy malaise in the space transportation area. As of the mid- to late 1980s, no real evidence of enlightened policy changes existed. However, the 1990s are ushering in an era of change at NASA and in the US space community. More and more calls are pursuing the development of space ships that will have much lower cost than current rockets and spacecraft. Innovative rocket

programs have been initiated to test new thinking in the areas of vehicle performance, maintainability, and operability. NASA is joining this movement. These
activities are in the realm of people most involved and interested in space
activities and are generally out of public view. Very few have heard anything in
the news about the push to begin development of economical space ships, but
it will likely be splashed across the headlines, once it becomes a reality.

No new space ships are expected before the year 2000. However, a review
of some of the activity that is steering NASA and the space industry in the
direction of cheap and assured access to space will give an idea of current
trends.

The National Commission on Space Report

Among the many things discussed in this vision of America's future in space
was space transportation. The Pioneering the Space Frontier plan outlined five
thrusts for the civilian space program in the future. The report stated that it was
necessary to achieve low-cost access to space not only for earth orbit, but for the
entire inner solar system as well.

In its plan, the National Commission on Space (NCOS) laid out an agenda
to open up the space frontier to more people and organizations. The first thing
the report identified was the need to encourage long-term space exploration
and development. "The two most significant contributions the US government
can make to opening the space frontier are to ensure continuity of launch
services and to reduce drastically transportation costs within the inner solar
system."[136] To accomplish these goals, the NCOS report made several suggestions.

One important idea involved a next-generation space ship designed so that
commercial practices could be applied to the vehicle to reduce costs. These new
ships should also be more automated, autonomous, and require fewer ground
personnel in their operations. An all-weather capability and the ability to
rapidly refuel were also deemed as desirable traits of new space ships.
Although space-related contractors have built most space vehicles, they have
been dictated in design and specifications to meet NASA's needs. The idea here
is to give space companies more leeway in their designs to accomplish the goal
of opening access to space. NASA can be included in this process by giving out
contracts that promote innovative thinking in the design and construction of
new spacecraft.

Also, the commission recommended that NASA funding be increased to
support space technology development. With this in mind, the report
recommended seven specific technology demonstration projects for NASA that
included advanced rockets and aerospace planes. NASA's budget for space
technology research has been minuscule for many years now and this needs to
be addressed by the US Congress. It is hard for NASA to excel in advancing the

state of space technology without an appropriate budget to support this goal. The NCOS report also identified three space transportation needs for year 2000. These include cargo transport to low earth orbit, passenger transport to and from low earth orbit, and round-trip transportation beyond low earth orbit. Currently, the shuttle is assigned to carry both people and cargo into space. The commission team thought it wise to separate these two functions. This would allow specialized development of these craft that could cut costs in space operations. Piloted ships could subsequently become smaller in the future, but be capable of carrying more people. Cargo ships could stay large with their mission of delivering goods to orbit and beyond. To open up the space frontier, the report suggested that cargo transport costs should only be about $200 per pound by the year 2000 (1986 dollars).

NASA Access to Space Study

Although much of the single-stage-to-orbit (SSTO) work originated within the Defense Department, NASA has been coming around and has begun to take a serious look at the possibilities of SSTO technologies and their potential benefit to the nation. In early 1993, NASA began a year-long look at the future of American access to space through the year 2030.

Before designing any kind of transportation vehicle, decisions had to be made about what the vehicle will do, how it will do it, and how often it will do it. With a space ship, these and other questions need to be addressed. What orbit will the ship be able to reach? How long does it take to inspect and outfit a ship between missions? Will it launch from a traditional launch pad or will it take off like a plane? Will it land like a plane or descend vertically? Will it be powered by rockets or scramjets? The authors of this study examined what a shuttle follow-on should be able to do and selected four broad criteria: 1) space transportation costs should be reduced by 50 percent; 2) crew safety should be improved by an order of magnitude; 3) vehicle reliability should exceed 98 percent; and 4) operability issues -- how easy it is to prepare the vehicle for missions -- should be vastly improved. The goals being aspired to are obvious. Future space ships need to operate with the frequency and reliability of airplanes and jets. This is the only way that space travel will ever hope to become accessible to a large number of people.

With these four goals in mind, the study team broke into three groups and considered different ways to reach their goals. The first team explored ways to keep the space shuttle and current rockets operating by upgrading the hardware in the vehicles. The second team looked to develop new rockets using conventional technology. The third team examined new reusable vehicles that incorporate advanced technology to achieve the study's goals.

The third study team decided to look at space flight differently than the first two teams. The overall strategy employed stressed flexible and reliable operations instead of a focus on performance by using advanced technology.

The team also stressed the need for a "culture change" in developing and operating a vehicle to maximize cost savings. This team, for example, deemed autonomous flight operations essential. This means that the ship's onboard computers should be able to monitor the ship's operation to a greater extent than current vehicles. The final design should also be able to fly at least 39 times a year and be capable of placing 25,000 pounds of payload into a 220 nautical mile circular orbit.

Designs should need only a one-time vehicle flight certification, off-line payload processing, the minimization of serial processing, a durable heat shield for reentry, and autonomous avionics. The shuttle, by comparison, needs detailed inspections after each mission. Also, some shuttle processing is hazardous (the handling of thruster chemicals, for one) and must be done while other work must wait. A heat shield system not subject to nicks and cracks would also eliminate the need for detailed tile inspections and replacements, such as that on the shuttle. The shuttle era is giving NASA some valuable insights into how to make a better space ship and this knowledge will help future designs. All the requirements above will minimize the need for constant scrutiny of the vehicle and will speed along the turnaround time between missions. The ability to fly, and fly often, will do much to make future space ships affordable.

This last design team also envisioned three design concepts that use advanced technology. The first is a rocket-powered SSTO ship that is powered by a Russian RD-704 class engine. This SSTO would take off vertically and land horizontally. A second idea included an SSTO that would not only use rockets, but also ramjets and scramjets. This ship would take off and land horizontally, such as an airplane and would possess many aircraft-like characteristics. The last design is a two-stage-to-orbit (TSTO) vehicle that also uses scramjets and rockets, but uses a large booster vehicle for the first part of the ascent before separating from a smaller orbiter that goes into space.

The rocket-powered SSTO became the final choice of the design team. They recommended to begin a five-year development plan that would cost $900 million. The estimated cost of developing this vehicle came to approximately $17 billion in Fiscal Year 1994 dollars. The resulting annual operations costs were, however, estimated to be the lowest of the three options at $1.4 billion per year.

This study points to a stark reality that policy makers need to understand. No significant reduction in the cost of going into space is going to happen with the appropriation of a small amount of money. The potential benefits of such a radical reduction in the cost of space flight is more than worth a mere $20 billion in development costs. This is a considerable amount in everyday terms, but it is not much compared with our federal government's budget. Our government has a history of helping people to open up the frontiers of our unexplored nation in its early years and it would be wise for our government to promote the opening of the space frontier. The return on the investment will more than dwarf the money spent to make space truly accessible to all people.

The Delta Clipper

The idea of low-cost space ships is nothing new. The problem is that neither the US government nor the aerospace industry have really made an effort to work on a new ship that could open up the space frontier to everyone. Recently, though, the very first steps towards achieving this goal were taken. One would expect such an initiative to originate from NASA or the enlightened policy of a president. However, the US Department of Defense introduced it, specifically, the now-renamed Strategic Defense Initiative Organization (SDIO).[137]

In 1983, President Ronald Reagan proposed the creation of a space-based missile defense system that could protect America from a surprise nuclear missile attack by the Soviet Union. This idea involved enormous amounts of hardware and money that would end up silently circling the globe serving as a global guardian against nuclear catastrophe. Buried within this idea was a low-cost, rapid response piloted space ship that would deliver much of the "Star Wars" technology to orbit. Reagan's "Star Wars" idea floundered in Congress. The space-based version of the defense shield eventually died. SDIO in 1989 commissioned a study that concluded that building an operational single-stage-to-orbit (SSTO) space ship was possible at present. This study then helped SDIO create the Single Stage Rocket Technology (SSRT) program that had the initial goal of building a low-cost and operable SSTO vehicle.

As in most spacecraft programs, a competition began in the aerospace industry to propose the best possible designs to meet the needs of SDIO's ship. Phase One of the program dealt with soliciting designs that were eventually submitted by McDonnell Douglas, General Dynamics, Boeing, and Rockwell International. Both McDonnell Douglas and General Dynamics submitted bids for vertical takeoff and landing vehicles. Boeing used its extensive aircraft knowledge and proposed a vehicle that would takeoff and land horizontally like an airplane. Rockwell used its space shuttle knowledge to submit a design for a vehicle that would launch vertically and land horizontally. SDIO selected McDonnell Douglas as the winner in August 1991 and awarded the company $59 million to conduct Phase Two of the SSRT project. McDonnell Douglas's proposal, the building and flight testing of the *Delta Clipper-Experimental* (DC-X), would take place over a two-year period. The DC-X was proposed as a one-third scale model of an operational SSTO and would test many of the requirements needed in a real SSTO. Quick turnaround times, aircraft-like maintenance, autonomous controls and checks, and a small ground crew were some of the goals of this flight test program.

McDonnell Douglas then set off to build its ship and prove its concept. The resulting vehicle looked something like a cone, but with a square base instead

The *Delta Clipper-Experimental* (DC-X) vehicle lifts off during a test flight (left) and ascends to 300 feet during another flight (right). (Courtesy McDonell Douglas Space Systems Co.)

of the expected circular features. The 42-foot-tall, 14-foot-wide at-the-base ship weighed 22,760 pounds empty and 41,630 pounds fuelled. DC-X had a mass fraction of 0.5 -- meaning that 50 percent of the ship's mass consisted of fuel -- and included a composite aeroshell, an aluminum frame and associated tanks, many off-the-shelf components, and an automated flight control system. From the beginning, this project was not meant to be normal. Many space projects require several years and even decades to get from the drawing board to the final frontier. The DC-X went from creation to flight testing in only two years. The designers in this program thought in terms of aircraft operations. A real space ship should be able to go out on a mission, come home, refuel, reload, take on a fresh crew, and launch back into space. DC-X would show that some of the goals could be accomplished.

Not designed to fly into space, the DC-X had a three-part flight test program that would show some of the capabilities needed for SSTO operations. The first two phases involved takeoff and ascent, a translational (sideways) movement while hovering in the air, and a landing. These two phases were to take the vehicle to successively higher altitudes up to 5,000 feet above the ground. Phase three of the program was even more aggressive. The DC-X would be lifted to 20,000 feet above the ground and pitched over in-flight so the vehicle could simulate a landing that would place the craft right where it took off. Controlling the vehicle while in-flight involved only three people. Tucked away inside a semi-trailer outfitted with an array of electronics gear, the makeshift Flight Operations Control Center (FOCC) included two flight crew members and a ground operations controller. Only seven people were needed for DC-X maintenance and about 25 people worked in various support operations. This is in contrast to the space shuttle, where thousands of people are involved in the upkeep and maintenance of the fleet. The analogy between the DC-X and shuttle is not very appropriate because DC-X is not a real SSTO vehicle, but the ability to reduce the number of people needed to maintain a space ship fleet will be vital to the future of space operations.

Once the vehicle and its equipment were ready, the program moved to NASA's White Sands Test Facility in New Mexico to check out the engines and systems of the DC-X. May and June of 1993 involved various engine firings of the four DC-X liquid oxygen/liquid hydrogen engines provided by Rockwell International. In one of the tests, two engine firings of the DC-X were conducted in only an eight hour time span. From then on, the vehicle was ready to take to the air. The DC-X first left Earth on August 18, 1993, for a hover flight test that took it to only 150 feet in altitude. The vehicle rose up to its altitude, moved sideways, and settled back down to Earth. With success in hand, McDonnell Douglas and SDIO pushed forward with their flight tests. A similar test took place on September 11, 1993. This 66 second test took DC-X 300 feet high, followed by a 350-foot lateral maneuver, and a landing. A third test took place on September 30, 1993, as the ship went up to 1,200 feet in altitude while going through the maneuvers of the previous tests.

Although everything went as good as anyone could have expected, storm clouds soon began forming over the DC-X test program. Two problems started to surface in the program: equipment failure and money shortages. The next test flight, scheduled for October 21, 1993, stopped 1.5 seconds after the engines began firing because sensors on the ship detected that one of the engines did not generate enough thrust. This flight, originally scheduled for October 20, and an October 23 flight, were to demonstrate that the DC-X could be flown twice in three days. Both missions were ultimately canceled due to a lack of money and the whole effort suddenly came to a screeching halt.

DC-X flights resumed in June 1994 and began with a successful flight to 2,600 feet in altitude on June 20. A week later, on June 27, another test flight did not go quite so well. A DC-X flight had to be aborted after only 17 seconds of flight when the shock wave from the hydrogen explosion ripped a hole into the side of the vehicle. Later it was found to have been caused by the ignition of gaseous hydrogen vented nearby into the atmosphere. The DC-X used its automatic abort mode to land safely, but the vehicle appeared to have been struck by a missile.

The next several months were then occupied with DC-X repairs and renewed political fights in Washington to obtain funding for SSTO research. In early 1995, the Advanced Research Projects Agency transferred $35 million of its congressionally-appropriated SSTO funding to the US Air Force Phillips Laboratory. This allowed work to begin on SSTO-related projects and for continuation of the DC-X flight test program. With this release of funding, McDonnell Douglas repaired the DC-X and the vehicle returned to White Sands on April 7, 1995. The sixth DC-X test flight quickly took to the skies on May 16, 1995, in a 123 second mission that saw the vehicle increase its performance compared to previous test flights. In this test, DC-X climbed skyward in a slanted ascent to an altitude of 4,350 feet, then moved horizontally until positioned over the landing pad and finished the flight with a normal landing.

Following another test flight in June, the DC-X began its eighth and final test flight on July 7, 1995. The main objective of this flight involved in-flight, powered rotations of the vehicle to show that a real DC-X-type SSTO can be properly maneuvered to permit a landing. A DC-X-type SSTO is envisioned to reenter the atmosphere nose-first and be able to pitch-over to a tail-first attitude in the lower atmosphere to allow a vertical landing. Rising to altitudes of more than 6,000 feet, the DC-X performed two rotations of about 180° each. The original program goal of having the DC-X perform a simulated landing as if it had reentered from earth orbit had to be dropped. Flight data showed the vehicle lacked the aerodynamic control to perform the maneuver. However, the in-flight rotations provided evidence that this style of vehicle could perform the required maneuvers to land as hoped. The DC-X landed at a faster speed than expected and suffered a crack in its aeroshell, but no serious damage resulted and this finished the vehicle's first series of test flights.

Shortly thereafter, NASA announced that the DC-X had been transferred to the space agency by the US Air Force to allow NASA to modify and test-fly the vehicle in support of its Reusable Launch Vehicle (RLV) technology program. The RLV program, which includes the design and possible development of two X-vehicles, the X-33 and X-34, has the overall goal of significantly reducing the cost of access to space. By achieving this goal, it is hoped that new space-based services will emerge with a subsequent positive impact on improving US economic competitiveness. Renamed the *Delta Clipper - Experimental Advanced* (DC-XA), NASA will use the knowledge gained from flying this vehicle to improve the design and performance of its X-vehicles. DC-XA modifications will include installation of a Russian-built aluminum-lithium liquid-oxygen tank, a lightweight graphite-epoxy liquid-hydrogen tank, a composite intertank and a fluid-to-gas converter assembly in the DC-XA flight reaction control system. It should be noted that the use of composite materials, as in the fuel tanks just listed, offers the strength of traditional metal fuel tanks but with a significant reduction in weight. NASA began test flights of the DC-XA in June 1996 in the New Mexico desert.

NASA's RLV technology program is a relatively new creation and is institutionalizing the idea of low-cost space access within the space agency. The X-33 is one of the prime efforts and is aimed at eventually producing an operational and low-cost SSTO for many different types of uses. Phase One of the project began on March 29, 1995, when NASA announced the selection of three aerospace companies who will conduct design studies that hopefully will lead to the construction of a prototype SSTO. The companies involved include the Lockheed Advanced Development Company, Palmdale, California; McDonnell Douglas Aerospace, Huntington Beach, California; and Rockwell International Corporation, Space Systems Division, Downey, California. NASA is providing these companies with about $7 million each in support of their work. The trio of companies will also individually match the NASA funding with money from their respective corporations. At the end of this 15-month effort, a decision will be made on whether to build one or more of the submitted SSTO designs and to conduct rigorous test flights. If Phase Two is initiated, Phase Three will involve a decision to start construction of an operational SSTO. If this happens, then significant reductions in space flight access will occur and the day will be much closer when individual citizens may be able to afford a flight off their home planet.

Chapter 6: Public Opinion and Involvement

So far you have been exposed to a variety of ideas, rationales, and opinions about the importance of space exploration and the need for its expansion. Now we will discuss the opinion of United States citizens about their nation's space endeavors. We live in a democracy and, hence, Americans at least strive for the ideal that they should have a say in the business of local, state, and federal government. More accurately our governments can be described as republican in nature. We elect our political leaders to carry out the majority wishes of the electorate.

In any government activity, public support for a program or policy is crucial to its survival. NASA and its civil space programs are no an exception to this rule. However, the space program and the general public have historically had very little direct contact with each other. Since the early days of the space age, the public has been relegated to the status of cheerleader for US space efforts. Television images of crowds watching rocket launches in the 1960s have only changed to crowds watching space shuttle launches in the 1990s. Average citizens really do not have much of a say in what America does in space.

This secondary role for the citizenry may be contributing to Americans' view about NASA and space efforts. The space community often describes public support for space as very broad, but very shallow. This suggests that there are few people who actively oppose space activities. However, support for space vanishes when NASA funding is pitted against that of veterans' medical needs, environmental protection programs, or other plans that are more relevant to the average person's needs. When forced to choose among important government initiatives, space is not at all near the top of the list.

What do you think of NASA? Space efforts in general? Particular space programs? Have you ever really thought about it? What will follow is a review of several scientific polls, conducted by a variety of organizations that regularly or irregularly ask Americans about the US space program. Most polls try to cull answers of support or opposition, but some probe deeper into people's feelings and attitudes about the final frontier.

The space surveys below have their variations, but they reveal that Americans are really not very supportive of their space program. Surprisingly enough, this has historically been the case since the 1970s. Survey data on public attitudes toward the space program in the 1960s are not included, but the following years paint a picture of doubt and uncertainty for America's presence in space.

The Rockwell Poll

Since 1978, the Rockwell International Corporation has commissioned irregular polls to gauge the general public's attitude toward NASA and the civil space program. Rockwell is a major space contractor and has been responsible for building the space shuttle orbiters and the main engines used in the shuttle program. According to one of its latest polls in 1992, the American public is largely satisfied with current space activities. This survey, conducted from January 30 to February 6, 1992, by the polling firm of Yankelovich, Skelly and White/Clancy Shulman, involved a random sample of 1,000 registered voters from across the US who were called for 15-minute interviews. The specific purpose of the study was "to assess support for the United States space program, specific space exploration initiatives, the benefits derived from space exploration, and program funding."[138]

This poll differed from the other we will examine, because it does not try to make the respondents compare space activities to other federal programs, e.g., housing, welfare, or defense. The survey probed people's opinions about different space activities. The poll's first goal, to gauge support for NASA, produced some interesting results. Although more than 75 percent of those surveyed approved America's civil space program, only 42 percent considered civil space activities were extremely or very important to the nation. It also seems that these measure of importance to the nation have declined since 1988. Just before the shuttle resumed flight operations in July 1988, 59 percent of those surveyed viewed the space program as important. Two years later in February 1990, the same question produced agreement from 49 percent. Clearly, there seems to be a growing view that federal spending on space is unimportant.

When individual space missions were analyzed, the same trend emerged. Compared to the July 1988 poll, specific initiatives, e.g., space stations, new planetary probes, and missions to the Moon and Mars all lost support. The bright spot in this otherwise cloudy picture was that the majority still supported such missions. The 1992 poll showed support of 71 percent for planetary exploration, 65 percent for construction of a space station, 57 percent liked the idea of building a lunar base, and 49 percent wanted people to go to Mars. The space activities that received the largest support were Earth-monitoring satellites (91 percent) and joint, or international, space missions (77 percent).

Space technology transfer, on the other hand, seems to be a subject for almost agreement. The respondents surveyed strongly supported the practical applications of space technology and the transfer of public technology into the private sector. Those surveyed strongly supported the use of space technology for scientific and medical discoveries (92 percent), environmental monitoring

(88 percent), keeping children interested in math, science and engineering (88 percent), and for the creation of new consumer products (80 percent).

As expected, this poll also had its share of questions about the space shuttle program. The respondents who thought "the space shuttle is a remarkable technological achievement" chimed in at a strong 92 percent, while those favoring abandoning the space shuttle for unpiloted rockets only scored 29 percent. Also, when asked what the US should do if another space shuttle and crew were lost in an accident, 73 percent thought that the shuttles should continue to fly.

The most revealing part of this poll, however, involved questions on basic information about the NASA budget and its relation to the overall federal budget. The respondents' lack of knowledge was evident as 43 percent did not know NASA's annual budget. Among respondents who offered a guess, "The average proportion of the federal budget [for NASA] is believed to be 13.4%. Lower socio-economic groups, including the less educated and lower income respondents, believe NASA receives an even higher amount, around 19%."[139] As discussed in Chapter 3, NASA has historically received about one percent of the federal budget since the end of the Apollo era. When respondents were informed of this, 65 percent agreed to increase the NASA budget to 1.5 percent of the overall federal budget. With the NASA budget of about $14 billion in the early 1990s, a 50 percent increase amounts to about $21 billion. These results seem to confirm the view that most Americans are unfamiliar with their space program. The poll also found that only 2 percent of the respondents thought they were extremely familiar with space activities while 30 percent thought that they were only slightly familiar. Another 24 percent admitted having very little knowledge of US space activities. These results suggest that there is a very small group of Americans who closely follow space activities, while the majority has only a fleeting knowledge of space endeavors.

The Gallup Poll

The Gallup polling firm is just one of the national organizations that regularly polls Americans about a wide variety of issues. The space program is no exception. A 1989 survey[140] attempted to provide some reasons for the declining support for the space program. Two of the identified factors were a concern about domestic issues and a reduction in US-Soviet hostilities and competition. The great Moon race was predicated on the need to beat the Soviets to the Moon. It was to show the world that democracy and free enterprise were superior to communism and a state-controlled economy. If this rivalry had not existed, it is doubtful that anyone would have gone to the Moon in the 1960s. Now that the Soviet Union no longer exists, the shift in space has been toward cooperation. This can be seen in shuttle flights to Russia's *Mir* space station and plans for an international space station at the end of the decade. The poll continued that "the improvement in relations between the two superpowers compared to the early 1960's has taken away a sense of urgency Americans once felt about being ahead in the space race."[141] The poll also suggested that "...the space program may be suffering from its association with the defense department at a time when four in ten Americans want to reduce US military spending."[142]

Demographically, the Gallup poll produced several distinctive trends that clearly showed likely supporters and opponents of space exploration. Among men and women, men are more likely to see the investment in space as worthwhile (51 percent to 35 percent). In the three Gallup age categories, people aged 30-49 supported space more than those aged 18-29 and 50 and over (50 percent to 43 percent to 35 percent, respectively). However, the most striking disparities are seen from racial, educational, and income breakdowns of the population.

This poll only distinguished race into white and non-white groups. This simple classification may hide some attitude differences among minority groups, but still reaches a startling conclusion. White respondents were significantly more inclined to support the investment in space compared to non-whites (46 percent to 22 percent). Respondents with a higher level of education are more inclined to support space activities. For example, only 26 percent of those without high school diploma supported space, while 66 percent of them thought the investment could have been better spent elsewhere. Conversely, a majority of college graduates (58 percent) supported space while 39 percent did not. Men and women college graduates clearly supported the US space investment (63 percent and 47 percent, respectively) compared to those without college degrees (42 percent and 28 percent, respectively).

The breakdown based on personal income in the survey is also revealing. As income increases, so does support for space. People with an annual income of $50,000 or more support space (63 percent) more than people with lesser incomes. There is a steady drop in support for space as income declines. Only 28 percent of those with incomes below $20,000 agreed that the space investment was worthwhile. A final category involved political party identi-fication. A slight majority of Republicans supported the space investment (52 percent) compared to minorities of Democrats (31 percent) and Independents (45 percent).

Although serious thought is required to explain these survey findings, it is clear that minorities, women, the lesser-educated and the poor in America have a decidedly negative view of space exploration.

Some likely reasons for the above trends can be suggested here. Since the beginning of the Space Age, the US civil space program has been dominated by white men. They were the first to go into space, to run mission control in Houston, and to be predominantly employed in the scientific and engineering jobs at NASA. That is still generally the case today, but efforts are being made to encourage women and minorities to move into the historically white male-dominated professions of science and engineering. This sense of exclusion by women and minorities may contribute to their hostility to space activities. Likewise, poor people might feel space missions are frivolous when they are barely able to make ends meet to raise their families. Some poor people may feel that federal money should be spent to help them instead of on space probes.

A resolution of these sharp divisions in attitude will not come about overnight. The adventure of the human migration into space urgently needs to be expanded so that it includes everyone. Presently, some people may feel that space is the province of government and big business and excludes the average citizen. This may lead to attitudes of indifference and the conclusion that space activities do not affect "real" people. We may generate more support for space by opening up access to space for everyone, by conducting projects, such as solar power satellites that can help all the people of Earth, and by educating people that space is not reserved for the rich and the elite.

Another example of Americans' lack of knowledge about their space program is also appropriate here. The crowning achievement of US space endeavors so far has to be the Apollo missions to the Moon. Apollo 11 landed on the Moon on July 20, 1969, and the lunar lander Eagle carried Neil Armstrong and Buzz Aldrin to the lunar surface. Armstrong became the first person to walk on another celestial body that day. When respondents were asked if they knew who had declared, "This is one small step for a man, one giant leap for mankind" and what event it was associated with, 59 percent identified Armstrong or the lunar landing. The other 41 percent, however, did not know or gave wrong answers. In a similar question about the first person to walk on the Moon, only 39 percent identified Armstrong, 24 percent gave wrong answers, and 37 percent did not know at all.

Two years later, in 1991, Gallup surveyed Americans again about the space program. One of the poll's main conclusions was damning. "Space exploration remains a low priority for most Americans in comparison with other government programs on which tax dollars are spent. Overall, more than half (56 percent) of all Americans think the money this country has invested in space research would have been better spent on programs, such as health care and education." This type of result questions whether Americans really want their country to explore space. There is always that small constituency in support of space activities, but the poll results suggest most Americans would be just as happy if planet Earth was the universe in its entirety.

The results of other Gallup poll questions were equally negative. Most respondents considered it unimportant for the US to be the first country to land a person on Mars. Some 64 percent found that goal "not too important" or "not at all important", compared to only 8 percent who thought it was "very important" and 27 percent who considered it "somewhat important." This last question can now be considered outdated, because it is increasingly apparent that large-scale space initiatives, such as a Mars mission, need to be international in character. The days of one nation dominating the space scene have passed. To optimally leverage human society's resources, large projects will have to involve global cooperation. The question about Mars is revealing, however, because it shows most people are not very interested in humans travelling to other planets.

Similarly, another 1991 Gallup poll question asked respondents about their preference between a space program that concentrated on unpiloted missions (e.g., planetary probes and environmental monitoring) and one that emphasized piloted missions (e.g., space shuttle). The results showed 39 percent supporting the unpiloted option and 49 percent the piloted option. Such questions are quite irrelevant, however, because any aggressive human exploration of the solar system will require robotic probes and automated equipment. It is not an either-or proposition.

Many space scientists have opposed human space flight because they think the funding should be made available to them for their research. They point to the lower costs of probes versus piloted spacecraft and argue that NASA has unfairly supported piloted programs over their research. A robotic spacecraft-only space program will surely cost less than piloted missions, but we will lose much more than we gain. We may save billions of dollars by scrapping the shuttle and any possibility of future space ships. We would, however, pay a higher price in limiting ourselves to our overpopulated, resource-depleted planet and by ignoring the rest of the universe.

Public support for space exploration can be placed into a historical perspective (see Tables 5 and 6, and Figure 6). Table 5 shows a consolidation of individual polls from the major network television stations of ABC, NBC, and CBS. Table 6 summarizes the results from the General Social Survey polls on space spending since 1973. The general question posed in all these polls is whether the US is "spending too much, too little, or about the right amount on space exploration programs?" Both tables reveal a great deal of ambivalence among Americans about the value of space exploration. The network news polls demonstrate consistent results in which close to 20 percent feel too little is spent on space, 35-40 percent believe too much is spent on space, and some 35 percent think spending is appropriate.

The General Social Survey data were collected over a longer period and are more indicative of public opinion trends. The early 1970s showed a public that clearly felt too much money was devoted to the space program. This period coincides with the Apollo program winding down, the US conquering the Moon, and declining federal commitment to space exploration. The percentage of people thought space expenditures were excessive began to decline in the late 1970s to a relatively stable level of about 40 percent. The percentage of respondents who thought spending was appropriate ranged from 30-45 percent over the years, and citizens who thought insufficient money was spent on America's space effort has lagged badly in the range of 8-18 percent.

TABLE 5

Network Television News Polls
Question: Are we spending too much, too little, or about the right amount on space exploration programs?

Year/Network	Too much	Right amount	Too little	No opinion
1981 - CBS	36%	37%	18%	9%
1981 - NBC	31%	38%	22%	9%
1982 - CBS	42%	27%	18%	13%
1986 - CBS	40%	37%	12%	11%
Jan. '88 - CBS	43%	30%	19%	8%
Oct. '88 - CBS	33%	37%	19%	11%

Source: Poll data from CBS and NBC

TABLE 6

General Social Survey (GSS) Polls
Question: "...are we spending too much, too little, or about the right amount on [the] space exploration program?"

Year	Too little	About right	Too much	Don't know
1973	7.45%	29.30%	58.40%	4.79%
1974	7.68%	27.50%	61.00%	3.84%
1975	7.38%	30.10%	58.10%	4.36%
1976	9.14%	27.95%	60.20%	2.67%
1977	10.10%	34.40%	49.61%	5.88%
1978	11.55%	35.00%	47.19%	6.27%
1980	18.00%	34.47%	39.10%	8.45%
1982	12.35%	41.24%	40.10%	6.31%
1983	13.82%	40.46%	39.65%	6.66%
1984	11.76%	42.60%	38.95%	6.69%
1985	11.45%	44.07%	40.48%	3.99%
1986	11.23%	43.29%	40.96%	4.52%
1988	17.69%	41.78%	34.26%	6.27%
1989	14.71%	44.01%	34.64%	6.64%
1990	11.13%	43.62%	39.02%	6.23%
1991	11.69%	42.46%	37.70%	8.20%

Source: Poll data from the General Social Survey (GSS)

NASA's 1992 Town Meetings

Opinion polls are generally the most efficient way to determine a population's views about a specific subject, but they are not the only way. In 1992, NASA decided to hold a series of meetings across the United States to attempt to gauge public opinion about their space program. Specifically, NASA wanted comments about its draft vision statement, opinions about current NASA programs, and ideas for the future course of the space agency.

NASA held town meetings at six universities in November and December 1992. Sites ranged from North Carolina State University in Raleigh, North Carolina and the University of Hartford in Connecticut, to the University of South Florida in Tampa and the University of Washington in Seattle. The other two sites were Indiana University/Purdue University in Indianapolis and California State University near Los Angeles. More than 4,500 people attended the six meetings and many more wrote letters to NASA because they were unable to attend. Of the attendees, more than half came from the aerospace industry, 12 percent were university students or faculty, and the remaining 38 percent called themselves "interested citizens." The attendees also showed interest in almost all areas of NASA's space program. While 51 percent expressed an interest in some aspect of human space flight (space station, human exploration, and space transportation), 17 percent favored space science, 13 percent preferred the educational aspects of space, 10 percent chose Earth science, and 9 percent listed aeronautics.

Before the public made comments or suggestions at each meeting, NASA administrator Daniel Goldin described some of NASA's activities, goals, and future plans. During these meetings, he and other NASA officials declared that NASA intended to support space science funding at about 20 percent of NASA's overall budget. Other ongoing activities were to try to reduce the cost of large programs, such as the space shuttle and space station. They also stated that aeronautics research would increase to hopefully provide technology to the US aviation industry and to allow it to compete better internationally. Also, another NASA goal was to fly a larger number of affordable, small planetary and science spacecraft to allow more missions to be flown.

One of the most revealing aspects of these town meetings, however, was the large number of ideas and suggestions people made to help the space agency. The most popular comments from the audience dealt with NASA's communicates with the public, the need for improved space transportation systems, educational activities, more effective distribution of NASA materials to teachers, and improved NASA's technology transfer and research efforts. About one-sixth of all comments dealt with more effective NASA promotion to the public. Audience ideas included the sponsoring of a Super Bowl half-time show or a NASA-run television show that would air on the major networks. Although the officials did not support these unorthodox ideas, they indicated

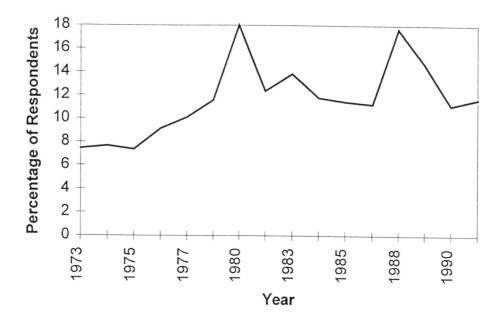

Figure 6. Respondents who think space program is insufficiently funded, 1973-1991.
Source: Poll data from the General Social Survey (GSS).

that NASA was increasing its distribution of press releases around the country. They are now sent to newspaper science and medical writers. NASA also improved its ability to send information to minority communities. Besides that, NASA has been working to foster greater public understanding of the science activities that occur on shuttle missions. Interactive television programming, improved radio programs, a public speaking program, and a new traveling exhibit are other activities to bring the public more into contact with NASA's space program.

The attendees also discussed two other popular topics: space transportation and fund raising. During these meetings, NASA officials indicated that they were considering three transportation options for the coming decades. They included upgrading and improving the current shuttle fleet, replacing the shuttle fleet by a new family of expendable launch vehicles, or pursuing development of advanced technology launchers, e.g., SSTOs. The officials added that the shuttle will launch the space station (although now with some help from the Russians) and that NASA did not plan to use any Russian rockets to launch US payloads. The idea of increased international cooperation was expressed. This attitude has been adapted by the space station program where Russia and the US are now heading up the project together.

The idea of building new and improved space vehicles struck a cord with many in the audience. For example, George Reeves of Raleigh wrote, "Transport cost reduction is the key to space access for all who want it. Offering every citizen a chance to vacation or work in space is a national goal that would have great appeal." Across the country in Washington, Robert Taylor declared, "Space should not be a spectator sport for the people of the world."

Many topics were discussed at these meetings, but many of the audiences' comments reflected the wide variety of opinions expressed by Americans in general. Both positive and negative opinions were expressed about space exploration. This is a good example of how difficult it is to set public policy in the face of a citizenry that is unclear about what it collectively wants. Take the comment by Ron Klein of Raleigh, who complained that he had grown "increasingly uncomfortable with NASA's message of exploration and national prestige at [a] time, when there are so many social problems. Are there long-term goals, such as exploration of Mars or the moon, that it might be appropriate to direct resources to solve social problems?" Dr. Paul Wolbers of Sarasota, Florida declared, "As long as there are so many Americans who can't afford health insurance, who don't have a home, who don't have sufficient [food] to eat, I think sinking money in NASA is a crime against the nation."

Many other comments, however, displayed different perspectives of how human movement into the solar system could benefit the species as a whole. Hal Horne of Culver City, California, wrote, "The underlying foundation of any activity that we undertake in space must be that humans from Earth will colonize and live permanently off-planet. Otherwise we'll still just keep crawling back to the cradle. And, who wants to live in a cradle?" Robert Rathbone of Tampa opined, "As I grow older, I do not wish to point to a small reddish dot in the sky and tell my grandchildren 'we have been there." [We do not need] another post card from space. I wish to someday point up to the heavens...and say 'we are there." Elizabeth Friedle of Yorktown, Indiana, stated, "The earth is crowded. Let's spread out. People should still be going out into space. Robots can't totally observe. What about feelings? If there is something scary, frightening, exciting, wonderful or beautiful that happens, how will a robot know?"

Ultimately, town meetings and public opinion polls can only give a snapshot of how a citizenry feel about something at a particular time. We need to focus on how to inform, interest, and educate people about the final frontier. There are groups to join, adventures to participate in, and ways to express yourself that can hasten the day when human society finally decides to spread out beyond Earth. At this point, I hope you are ready to get involved! What follows is how you can make your voice heard.

The National Space Society

The National Space Society (NSS) is one of the most active and influential space activist organizations in the country. Based in Washington, DC, NSS is the product of a 1987 merger of the National Space Institute and the L-5 Society. Both organizations brought to NSS differing strengths: L-5 formed as a grass-roots space education group that included local chapters from around the nation. NSI used its strength in Washington to help the space movement politically. Today, NSS continues to be an ever-louder voice for space on Capitol Hill and across the country. Its membership base, which extends to many different countries around the world, stands at about 27,000. In addition, NSS has continued the L-5 legacy of chapters by supporting the creation and local activism of dozens of chapters. They are active in their communities to inform people about the merits of space exploration.

Overall, NSS is strongly committed to expanding humanities ability to live and work throughout the solar system and to work in support of the creation of a spacefaring civilization. To make our species spacefaring, NSS -- as an organization -- has adopted eight broad goals. These include promoting space exploration, research, development, and habitation; providing broad and visible public support for space; encouraging the study of math and science by creating new space employment opportunities; seeking maximum support for space technology research; providing public forums to exchange information; informing the public on space and space technologies; and recognizing and honoring of significant contributions to the space program. As far as space policy is concerned, NSS supports a wide variety of activities it feels advance the day when access to space will be available to all. For example, NSS supports the development of routine and low-cost access to space. Beyond that, NSS advocates NASA's space station and Mission to Planet Earth programs, a permanent human return to the Moon, the human exploration of Mars, commercial space development, and deep space exploration.

This is all well and good, you might say, but what is in it for you? Unfortunately, many non-profit and voluntary organizations do not try to do much beyond increasing their membership and annual membership dues. NSS does rely on its membership dues for its operating budget, but this group gives you a wonderful chance to be involved. You can start a chapter, attend or start a conference, write your congress person, invite speakers to your area to give a lecture, arrange a public display in a shopping mall to promote Spaceweek, or do any number of things. All that is needed are your initiative and drive to spread the word of space.

As an aside, I have been actively involved with NSS for several years now. It is amazing how much work a few people can accomplish when they are truly committed to their goal. From 1989 to 1992, I served as the president of a local NSS chapter at Iowa State University in Ames, Iowa. Iowa is definitely not the focus for US space activities, but I helped to revive an old L-5 chapter. The revived chapter limped along for the first couple of years, but in that period we managed to invite NASA astronaut Bryan O'Connor to Iowa State University for a talk.

Taking a bigger step, our chapter, the Iowa State Space Society (ISSS), sponsored a trip in which seven ISU students spent a week visiting NASA's Marshall Space Flight Center and Kennedy Space Center in 1990. During these early years, we also exhibited public displays on the ISU campus and in a local mall to publicize our cause. All these activities were carried out by chapters members who were volunteering their time and effort.

Our biggest project started in the fall of 1990 when four ISSS members traveled to Ohio to attend the Midwest Space Development Conference (MSDC). When they returned to Iowa, they immediately suggested that we should organize a conference of our own. This was a big idea, since we were all a group of college students with limited time and money. However, everyone accepted the idea.

By the summer of 1991, planning for the first Mid-Continent Space Development Conference (MCSDC) was well underway. Led by less than a dozen dedicated space supporters, the MCSDC took place on February 21, 1992. This almost three-day event included 25 speakers and more than 200 attendees. We had speakers from NASA, NSS, the aerospace industry, and other non-profit space-related groups. The program included sessions for teachers on how to teach space to their students and programs on activism to promote space. Other speakers discussed Mars, the Moon, solar power satellites, asteroids, and almost every other space subject. It is hard to describe the dedication of the people who made this project a success. The MCSDC has now become an annual conference in Ames and continues to make a name for itself in NSS. Many people do not have the time or inclination to do volunteer work for an organization. However, if you want the chance to make a difference, to learn about and inform people about the incredible promise of space, and to feel a sense of accomplishment, then NSS is the place for you.

Nationally, NSS conducts a wide range of functions to promote space exploration. Every year NSS sponsors an international space development conference that attracts people from around the world to hear and learn from the leaders of the space community. It is a time for networking, learning, and becoming active with colleagues from all over the US and abroad. Other NSS chapters also host a variety of regional conferences across the country. In addition, NSS works closely with two other groups -- *SpaceCause* and *SpacePac*. *SpaceCause* is dedicated to influencing space-related legislation in Congress, while *SpacePac* serves to provide contributions to politicians who are supportive of space exploration. If you are more interested in politics, then these groups may be for you. There really is no limit to what you can do or accomplish.

US Space Camp

There are many educational avenues that can be taken for both children and adults. One of the most popular programs for kids is the US Space Camp that has facilities in both Huntsville, Alabama, and near Kennedy Space Center, Florida. Space Camp conducts simulated shuttle and jet fighter missions for kids and adults. Located nearby NASA's Marshall Space Flight Center in Huntsville, the Alabama Space and Rocket Center consists of a sprawling area that includes a space museum, the Space Camp area, a rocket park with dozens of rockets used since the beginning of the space program, and a full-size replica of a space shuttle with attached external tank and solid rocket boosters. People do not just come here for Space Camp because of the huge variety of space memorabilia that is available for people to explore.

Since the beginning of Space Camp in the early 1980s, more than 140,000 people have graduated from the Camp's growing assortment of programs. The mission of Space Camp is simple. One goal is to use the excitement of the space program and military aviation to motivate young people toward greater achievement in math and science studies. Another goal is to promote the profession of teaching. The programs offered can be divided into three broad categories: programs for kids from fourth grade to twelfth grade, programs for teachers, and programs for adults. The link between the programs is their ability to expose children and adults from various backgrounds to the space program and to excite them to the possibilities for human society, if we decide to move into the cosmos.

Space camp for kids is divided into three areas based on age: US Space Camp for children in grades 4-6; US Space Academy, Level 1, for those in grades 7-9; and US Space Academy, Level 2, for teenagers in grades 10-12. All the programs are held at the Huntsville facility. There is also a US Space Camp for children in grades 4-7 that is located at the US Astronaut Hall Fame in Titusville, Florida.

Many of the programs have basic similarities, but the difficulty level increases with participant age. For example, the Alabama Space Camp program consists of a five-day program that includes classroom instruction, simulated astronaut training, and interactive lessons. During their adventure, the children learn about rocket propulsion and go on a tour of NASA's Marshall Space Flight Center and the Alabama Space and Rocket Center. In preparation for their eventual shuttle mission, the kids build and launch their model rockets and taste-test astronaut food, such as the dehydrated Neapolitan ice cream. Astronaut training always involves a variety of simulators and the kids use devices, such as the 5DF (Five Degrees of Freedom) machine that lets the user feel like a spacewalking astronaut. The kids also experience the multi-axis trainer that spins along three axes to simulate an out-of-control space ship. Another enjoyable device, the 1/6-th gravity chair, allows children to simulate a walk on the Moon. All this activity builds up in preparation for a simulated shuttle mission. Two missions take place so that each participant can be a member of the shuttle crew on one mission and a member of mission control on the other flight. While learning about the intricacies of space flight and the shuttle, children also begin to learn the value of teamwork. This is a lifelong skill that can be used by anybody. It is emphasized as children attempt to pilot their shuttle on its mission. The same general program offered here is also available at the Florida site.

The next level up is for junior high school students in grades 7-9. This five-day program began in 1984 and involves a more intensive study of the shuttle and space environment than that offered by the program for the younger campers. The kids get to learn more about the history of the US space program. They are even graced by the presence of members of Werner Von Braun's original rocket team who helped get America into space in the 1950s and the early 1960s. These long-time rocket experts talk about rocket propulsion and the science behind it.

As with any important upcoming mission, the kids exercise every morning and go through the simulations necessary to prepare for their missions. Among other events, the campers use a pool to simulate the microgravity environment of space. They even get to build three types of satellites in a "clean room." This is an ultra-clean room in which satellites are built. This sterile environment protects the satellite components from becoming clogged with dust or other debris. The big buildup leads to the simulated shuttle missions and the campers are initially assigned to be in the shuttle, on a space station, or in mission control. The mission takes place on a variety of hardware that includes a shuttle cockpit simulator, a full-sized Spacelab module, in a space station, and in a mission control center. Besides learning how to fly in space and conduct experiments, the kids learn about different kinds of space technology and learn about the types of careers needed to work for NASA and in the aerospace industry.

For high school students, the US Space Academy, Level 2, program gives them the chance to begin to seriously delve into their options for a future career. This eight-day program does an excellent job at mixing classroom learning with hands-on activities. The curriculum involved is much more focused and intense and the students who attend earn one semester hour of college freshman-level science credit from the University of Alabama in Huntsville. When at the Space Academy, participants are allowed to pick between one of three tracks that enables each student to focus on his or her area of interest. In the engineering track, campers serve as mission specialists on the shuttle or as space station specialists on a station. Scuba instruction is provided so students can sharpen their spacewalking skills. Other areas of instruction include robotics, optics, and different engineering fields. In the aerospace track, students with a desire to fly can learn what it takes to be a shuttle commander, shuttle pilot, or an aviation professional. Flight simulation training is offered at the Space Camp's Aviation Challenge Complex where campers also learn how to fly the shuttle. Typical study topics for this track include celestial navigation, orbital mechanics, and space piloting. For the scientists in the group, the technology track allows students to become payload specialists. These professionals are assigned to design and conduct experiments both on the shuttle and on the space station. Some of the experiments include gene mapping, human physiology testing, and soil and water analysis for contaminants. The diverse position of payload specialist also requires students to learn about emergency medical procedures, astrophysics, and materials processing experiments.

Besides the space-oriented programs already described, the camps also offer programs that focus on aviation for both junior and senior high school students. These five-day programs teach participants the ins and outs of military pilot training and what it takes to have the "right stuff." All the training involved builds up to an exciting air-to-air combat mission that pits students against each other in a simulated battle for survival.

Space Camp recognized years ago that all the fun should not be restricted to kids. They therefore instituted a series of programs geared towards teachers and adults interested in space exploration. There are two programs available in Alabama that introduce educators to curriculum materials and learning techniques to motivate students in the fields of math and science. Along with conducting shuttle missions as the kids do, teachers are informed about technology developments in the space industry, they listen to lectures, and receive classroom activities they can take home. For adults solely interested in the final frontier, there is the Adult Space Academy, Levels 1 and 2, that introduces non-aerospace professionals to the excitement of the space program. Learning to fly a space shuttle, do a spacewalk, or conduct experiments in a space station can be just as exciting for adults as it is for kids.

The Challenger Center

The Challenger Center for Space Science Education was founded by the families of the Challenger 51-L crew who perished in their journey to space on January 28, 1986. It is dedicated to bring the excitement of space to children all across America. The Challenger Center's objective is to promote mathematics and science education to school students, by using space as a motivating theme. The non-profit Challenger Center places an emphasis on hands-on activities, cooperative teamwork, and the encouragement of female and minority participation in their programs. Challenger Center focuses on four primary activities to accomplish its mission. It promotes the building and operation of Challenger Learning Centers across the country, conducts teacher workshops and conferences, and develops classroom project activities and curricula.

The Challenger Learning Centers are space simulators that include a mission control and a space station. The students conduct a variety of missions:

they may land on the Moon, go to Mars, or send a space probe to search for life on another planet. The simulators show students the value of working together as a team and handling unexpected problems. While assigned to a space station, a participant may build a space probe, identify and solve the life-threatening problem of an oxygen shortage, study plant biology, handle hazardous materials, or find the most suitable landing site for their space ship.

Twenty-five Learning Centers are located in museums, science centers, schools, and other facilities. They provide access for students who otherwise would not be able to participate in this type of activity. More than 200,000 middle school students attend Learning Centers annually, half of which are minorities. Missions are conducted in English, Spanish, Chinese, and French. Some Learning Centers are modified for use by mentally and physically impaired students. Challenger Center feels that it is important to provide access to as many children as possible in its efforts to encourage students to reach for their potential and their dreams.

Besides the simulators, the Challenger Center is also busy conducting workshops for teachers and students. About 2,000 teachers attend Challenger Center workshops every year. In 1991, "Suited for Space" served as the theme of a teleconference that included NASA astronaut Frederick Gregory. In addition, Challenger Center sponsors classroom projects that allow students from around the world to participate in year-long space missions to address the challenges of problem-solving in multicultural and multilingual situations.

One example of these classroom projects is "Marsville: The Cosmic Village." The goal of the Marsville project is to build a multi-national settlement on Mars. It starts out in the classroom where a designated mission team of 5-7 students and a teacher team leader work to develop a part of the Martian outpost. The team has a choice of designing and building the outpost's air supply, communications, food production and delivery, recreation, temperature control, transportation, waste management, or water supply systems. Once it has completed this task, a mission team will join up with two other mission teams from other schools. These three teams will comprise a habitat crew and are then jointly assigned to design an inflatable habitat for use on Mars. One project requirement is that the teams must learn to communicate with each other through electronic or written methods -- no in-person or phone contact is allowed. The summit of this project is Link-Up Day where the teams finally meet for the first time and build their systems and habitat for the outpost. This type of activity is invaluable in promoting self-confidence, teamwork, and communications skills in the accomplishment of a major project. By realizing that they can contribute to a team and accomplish a worthwhile goal, students hopefully leave these types of projects with goals and aspirations they may not have had before.

In a similar innovative move, Challenger Center has spearheaded the creation of "Mars City Alpha™". Mars City Alpha transports students to a classroom in the year 2043 where they are assigned to develop the first international human settlement on Mars. Working in the classroom as five independent science teams, the students research, develop, design and build a five-unit human habitat for twenty-five crew members on Mars. Targeting students in grades 5 through 8, the educational program emphasizes experiential learning based on the Challenger Center Instructional Model. It includes academic content, teamwork, creative and critical problem-solving and responsible decision-making. An indication of the quality of this organization's work is the Learning® Magazine's 1994 Teachers' Choice Awards Mars City Alpha received in recognition for a "practical, hands-on tool" for the classroom. It was developed under a grant from the US Department of Education Dwight D. Eisenhower National Program for Mathematics and Science Education.

The Challenger Center continues its work and is expanding. As of September 1995, there were 25 operational Learning Centers in the US and Canada. That number has since grown substantially. Overall, Challenger Center educational programs have reached more than 1.5 million students and 45,000 teachers nationwide. To give you a more local flavor of a Learning Center and its activities, I include an article I wrote in the summer of 1993. It describes a Learning Center's activities located at the Science Center of Iowa in Des Moines.

Challenger Learning Center in Des Moines Opens the Path to Space[143]

Back in 1986, our nation was shocked and dismayed as the space shuttle *Challenger* exploded in mid-air about one minute into its launch on a cold January morning. Reaction to this tragedy included a mix of grief, tears, and anger. However, some people treated that event as a chance to start over and build a legacy that honored the crew and mission of STS-51L. From those ashes of despair the Challenger Center for Space Science Education was born.

One of the main missions of the Challenger Center for Space Science Education, which is headquartered in Virginia, is to provide facilities called Challenger Learning Centers that give children a chance to experience a simulated space mission. In 1992, with the help of Kirk Brocker and others at the Science Center of Iowa in Des Moines, Iowa's only learning center opened its doors to Iowans. Located at the Science Center, the new facility has given more than 10,000 students the chance to go on a mission to outer space. From the beginning, the Iowa Space Grant Consortium (ISGC) has developed this project and made it a reality.

In the past two years, the scope and breadth of activities at the center have grown by leaps and bounds. From school programs and day camps to mini-flights and Marsville, people of all ages have opportunities to get involved. A new program called the Advanced Space Camp took place June 20-25, 1993. Eighteen children from sixth through eighth grades participated in the residential sleep-over program. "In June, the students went to Camp Dodge and underwent compass training", said Debbie Curry, who is the director of education at the Science Center and who played a significant role in getting the learning center up and running. "They also went to Saylorville Lake to try to find a suitable landing site for their mission. The students looked at four possible sites and they had to study the geology of the area, test the water, and do other investigations to find the best site."

Other activities included leadership training, model rocketry, and microgravity experiments in a YMCA pool. By paying for the living expenses of the campers, the ISGC promote the success of this new venture.

During the academic year, however, the school program is a two-hour mission: one hour in mission control and the other hour is spent in the space station. The teachers who accompany their classes must complete a one-day preparatory class to learn what to do and what will happen. The theme of each mission changes yearly. In 1992 students explored Halley's Comet, and this year [1993] they are going on a mission to the Moon. Next year's [1994] mission will include a journey to Mars.

According to Curry, the other programs at the learning center appeal to a much broader audience and allow the general public to get involved. "The weekend camp-in involves an overnight stay and the Saturday program, which lasts four hours, is popular with school and church groups", said Curry. "The mini-flight program is for the general public. Twenty people conduct a one-half hour mission in the space station. This program is for people of all ages."

In addition, a corporate program designed for company employees started last fall [1992]. This activity is geared at improving teamwork and communication skills in the setting of a space simulation.

Teachers from around the state also can get involved by attending any of the eight aerospace education workshops -- two of which are funded by the ISGC.

Looking to the future, the Challenger Center is initiating a national program called Marsville. Its goal is to have teams of students work together to build a small town that might exist on the surface of Mars. Within a state several different teams are working on the project. Each team works on a particular aspect of the town, such as life support systems, transportation, or other vital services. All this work culminates on "Link-Up Day", when the teams come together and build their habitats.

"A test project took place in Iowa on May 15, 1993, with nine pilot teams that built three habitats", noted Curry. "There were 52 children that participated on Link-Up Day. We wanted a varied group of children involved, and we had youngsters from 4-H, the Girl Scouts, home-schoolers, elementary education students, and children from a handicapped facility. The youngsters and the parents both gave the project high marks."

The Challenger Center works through Iowa's premier space simulator to promote a better future for the human race. The unique bond that ties together the vision of many people is the knowledge that the future is in space.

International Space University

For those interested in a more intense educational experience at the graduate school level, attendance at the International Space University (ISU) is recommended. Founded in April 1987 by Todd Hawley, Peter Diamandis, and Bob Richards, ISU is a growing institution that teaches graduate students about many different aspects of space activities. The guiding purpose of ISU is the promotion of the peaceful exploration and development of space. The upstart university has conducted a 10-week summer session at various locations around the world since 1988. This has led to the conferring of master's degrees to the students who have attended. In 1993, ISU selected Strasbourg, France, as the location for its permanent central campus system. ISU also plans to operate affiliate campuses and advanced campuses around the world. The advanced campuses are to be located at worldwide centers of space excellence that specialize in various areas.

To qualify for admission to ISU, students must have a bachelor's degree from an accredited university and should have some familiarity with the subjects taught at ISU. These include space architecture, space business and management, space policy and law, space resources and manufacturing, space engineering, space humanities, space informatics, satellite applications, space life sciences, and space physical sciences. Students are recommended to minimally have acceptance to a graduate-level academic program or to have a master's or doctoral degree. Alternatively, students should have at least two years of professional experience in government, industry, or academia after earning a bachelor's degree. Furthermore, ISU students must be proficient in speaking English and native English-speaking students must demonstrate proficiency in a second language. Clearly, this is a serious organization that wants to have a long-term impact on humanities movement into space. It is also a program where only the seriously motivated need apply.

Once at the summer session, the 10-week course includes 153 hours of lectures and instruction and 126 hours of design project work. The three-part program involves core lectures, advanced lectures, and the design projects. The core lectures introduce students to all the areas of study at ISU and the follow-on advanced lectures go into greater detail in each subject area. The culmination of the work is a design project that gives students a chance to design and work on a possible real-world project. This in-depth effort involves the formation of teams who examine every technical and non-technical aspect of the project. An examination of past projects shows their breadth and diversity. In 1989 at the Universite' Louis Pasteur in Strasbourg, students worked on the design of an International Lunar Polar Orbiter and a Variable Gravity Research Facility. The 1993 students at the University of Alabama in Huntsville worked on three separate tasks: a disaster prevention and warning system, a crew rescue and escape concept, and a lunar-based observatory. As described earlier in Chapter

4, the 1990 ISU summer session in Toronto had students working on the International Asteroid Mission. Students in that summer session also worked on the International Program for Earth Observations.

ISU also has short- and long-terms goals planned for its future. The university began offering a year-long Master's in Space Studies (MSS) program at its Strasbourg campus in the 1995-96 academic year. Long-range goals have ISU students conducting real research and educational activities in space. If you are serious about pursuing a career in any field related to space exploration, ISU is a program that you should consider. By 1993, more than 600 students from 47 nations have attended the summer sessions at ISU. With the start of permanent operations at Strasbourg, more and more people will have the opportunity to pursue their dreams in space. It is to be hoped that an increasing number of people around the world will realize that moving out into the universe is not a luxury for human society to pursue, but a necessity.

For more information:

Space Camp
1-800-63 SPACE
Conducts space and aviation simulations throughout the year for both children and adults.

The Planetary Society
65 N. Catalina Avenue
Pasadena, CA 91106
(818)-793-5100
Description: A public interest organization that emphasizes in planetary exploration and related activities.

National Space Society
922 Pennsylvania Avenue, SE
Washington, DC 20003-2140
(202)-543-1900
A diverse and broadly based space activist organization that supports aggressive public and private space exploration activities for the eventual establishment of a spacefaring civilization.

Space News
6883 Commercial Drive
Springfield, VA 22159
(703)-658-8400
A weekly trade newspaper that covers all aspects of the space industry.

Final Frontier
PO Box 16179
North Hollywood, CA 91615-6179
(818)-760-8983
Produces a bimonthly general interest space magazine of the same name. Also operates the Space Explorers Network -- an effort to bring together space enthusiasts from around the country.

Space Studies Institute
PO Box 82
Princeton, NJ 08542
A membership organization that promotes space research in fields, such as solar power satellites, electromagnetic railguns, asteroids, and much more.

Challenger Center for Space Science Education
1055 North Fairfax Street, Suite 100
Alexandria, VA 22314
(703)-683-9740
Works to encourage more math and science education in the public schools. Conducts teacher workshops, teleconferences, and promotes the building of the Challenger Learning Center space simulators across the country.

Young Astronaut Council
PO Box 65432
Washington, DC 20036

NASA Teacher Resource Center
US Space & Rocket Center
NASA TRC for MSFC
Huntsville, AL 35807
(205)-544-5812

HED Foundation
c/o Sara Ann Moody
PO Box 9421
Hampton, VA 23670-0065
(804)-826-0065

Bibliography

1. Chapman Research Group, Inc., An Exploration of Benefits from NASA "Spinoff" (Littleton, Co: Chapman Research Group, Inc., 1989), 3.
2. National Aeronautics and Space Administration, *Spinoff* 1981, Washington, D.C.: GPO, 1981, 65.
3. Ibid.
4. National Aeronautics and Space Administration, *Spinoff* 1982, Washington, D.C.: GPO, 1982, 98-99.
5. National Aeronautics and Space Administration, *Spinoff* 1986, Washington, D.C.: GPO, 1986, 54-56.
6. Ibid., 50-53.
7. Ibid., 51.
8. Ibid.
9. Ibid., 52.
10. National Aeronautics and Space Administration, *Spinoff* 1982, Washington, D.C.: GPO, 110-111.
11. National Aeronautics and Space Administration, *Spinoff* 1985, Washington, D.C.: GPO, 72.
12. National Aeronautics and Space Administration, *Spinoff* 1990, Washington, D.C.: GPO, 84-85.
13. National Aeronautics and Space Administration, *Spinoff* 1982, Washington, D.C.: GPO, 74-75.
14. National Aeronautics and Space Administration, *Spinoff* 1988, Washington, D.C.: GPO, 67.
15. National Aeronautics and Space Administration, *Spinoff* 1987, Washington, D.C.: GPO, 76-77.
16. National Aeronautics and Space Administration, *Spinoff* 1986, Washington, D.C.: GPO, 62..
17. Ibid., 63.
18. National Aeronautics and Space Administration, *Spinoff* 1990, Washington, D.C.: GPO, 76-77.
19. Ibid.
20. Ibid.
21. National Aeronautics and Space Administration, *Spinoff* 1986, Washington, D.C.: GPO, 84-85.
22. National Aeronautics and Space Administration, *Spinoff* 1991, Washington, D.C.: GPO, 98-99.
23.. National Aeronautics and Space Administration, *Spinoff* 1984, Washington, D.C.: GPO, 88.
24. National Aeronautics and Space Administration, *Spinoff* 1987, Washington, D.C.: GPO, 68.
25. Ibid., 68-69.
26. This segment is reprinted with only minor edits from a letter written by John Zipay to me dated December 17, 1992.
27. National Aeronautics and Space Administration, *Spinoff* 1989, Washington, D.C.: GPO, 56-63.
28. These testimonials have been reprinted with the permission of Sara Ann Moody.
29. Reprints courtesy of *Space News*. Copyright by Army Times Publishing Company, Springfield, Virginia.
30. Renee Saunders, "Satellite Imagery Confirms Source of Wetland Pollution", *Space News*, 29 April-5 May 1991, 25.
31. SPOT is a series of remote sensing satellites designed and built by the government of France.
32. Landsat is a series of U.S. remote sensing satellites that are owned by the U.S. government, but are operated and marketed privately.
33. Renee Saunders, "Imagery Lends Muscle to Yearly War Against Gypsy Moths," *Space News*, 27 May-2 June 1991, 26.
34. Renee Saunders, "Satellites Used to Trace Plutonium Runoff," *Space News*, 5-18 August 1991, 24.
35. Debra Polsky, "Imagery Aids Wildlife Study," *Space News*, 11-18 November 1991, 18.
36. Robina Riccitiello, "Satellites Used for U.N. Cambodian Refugee Relocation," *Space News*, 3-9 February 1992, 6, 8.
37. Debra Polsky Werner, "Colorado River's Irrigation Value Surveyed Via Satellite," *Space News*, 18-24 July 1994, 22.

38. Debra Polsky Werner, "Scientists Use Landsat to Hunt Ticks in Northeast U.S.," *Space News*, 15-28 August 1994, 16.

39. Andrew Lawler, "Chinese Use Japanese Satellite to Predict Earthquakes," *Space News*, 28 February-6 March 1994, 16.

40. Liz Tucci, "Imagery Helping to Determine Rain Forest Depletion Rate," *Space News*, 14-20 March 1994, 16.

41. William Boyer, "Shuttle Radar Sensor Used to Study Gorillas in the Mist," *Space News*, 20-26 June 1994, 10.

42. This is NASA press release 95-12, distributed to the media on February 7, 1995.

43. A megaparsec equals 3.26 million light years.

44. P.O. Lagage and E. Pantin, *Nature*, 369, 628-630, 1994.

45. For comparison to our solar system, Earth is 1 AU from the sun and Jupiter is 5 AU from the sun.

46. Wolszczan, Alexander, *Science*, 264, 538-542, 1994.

47. Mayor, Michel and Queloz, Didier, *Nature*, 378, 355-359, 1995.

48. "Factinos," *The Planetary Report*, May/June 1996, 20-21.

49. Congress appropriated $13.9 billion for NASA in FY 1996. The Clinton White House and Congress are planning to drastically reduce NASA's budget to around $11 billion by 2000. These moves could effectively kill space exploration for decades to come.

50. Budget of the United States Government, Fiscal Year 1991.

51. National Aeronautics and Space Administration, Social Sciences and Space Exploration, Washington, D.C.: GPO, 1984.

52. Ibid., 26.

53. Ibid.

54. Ibid.

55. Management Information Services, Inc., The Private Sector Economic and Employment Benefits to the Nation and to each State of Proposed FY 1990 NASA Procurement Expenditures, (Washington, D.C.: Management Information Services, Inc., 1989).

56. Ibid., v.

57. Ibid., 20.

58. U.S. Department of Commerce, Economics and Statistics Administration, Statistical Abstract of the United States 1993, (Washington, D.C.: GPO, 1993), 408.

59. Ibid., 182.

60. Ibid.

61. ref. 55, p. 46.

62. Ibid., 43.

63. Ibid., 42.

64. Chapman Research Group, Inc., *An Exploration of Benefits* from NASA *"Spinoff,"* (Littleton, Co: Chapman Research Group, Inc., 1989).

65. Ibid., 27.

66. Ibid., 28.

67. US Department of Commerce, International Trade Administration, *US Industrial Outlook*, 1994.

68. "U.S. Commercial Space Revenues Projected to Hit $5 Billion in 1992," *Aviation Week and Space Technology*, 29 June 1992, 68-69.

69. As a way to generate more money in the face of decreasing budgets, NASA is considering flying select military and commercial payloads again. Authorization for this would have to come from the president.

70. "New Launchers Elbow into Crowded Market," *Space News*, 7-13 March 1994, 8.

71. Andrew Lawler, "Successful H-2 Lofts Japan Into World Market," *Space News*, 14-20 February 1994, 1.

72. Andrew Lawler, "Culture, Fish Shape Tanegashima Operations," *Space News*, 14-20 February 1994, 11.

73. Daniel J. Marcus, "Space Unit Buoys McDonnell Douglas' Profits," *Space News*, 5-18 August

1991, 26.

74. William Boyer, "General Dynamics Stands Firm on Launch Business," *Space News*, 25-31 March 1991, 8.
75. 1991 Shareholders Report, General Dynamics, 29.
76. Daniel J. Marcus, "Debris in Atlas 1 Centaur Stage Likely Cause of Launch Failure,"*Space News*, 8-14 July 1991, 21.
77. William Harwood, "General Dynamics Delays Atlas Flight to Finish Tests," *Space News*, 8-14 March 1993, 21.
78. Theresa Foley, "GD Negotiating Sale of its Atlas Division to Martin Marietta," *Space News*, 11-17 October 1993, 6.
79. Ibid.
80. Patrick Seitz, "GD Works to Repair Atlas' Image," *Space News*, 31 May-6 June 1993, 4.
81. Lon Rains and William Boyer, "Pegasus Orbits Two Satellites,"*Space News*, 9-15 April 1990, 21.
82. Patrick Seitz, "High Hopes for Follow-on Pegasus," *Space News*, 6-12 December 1993, 17.
83. Harold L. Levin, "The Earth Through Time," 4th edn., (Orlando: Saunders College Publishing, 1994), 258, 262.
84. The TOMS instrument onboard Meteor-3 ceased functioning in December 1994.
85. Richard Wagner, "The Ozone, Lost Again," *Ad Astra*, January/February 1994, 7.
86. Lester R. Brown, Hal Kane and Ed Ayres, "Vital Signs," 1993, (New York: W.W. Norton & Co.), 67.
87. Ibid, 68.
88. Leonard David, "Save the Planet from Space," *Ad Astra*, May 1992, 16.
89. Douglas Fulmer, "Turning Up the Heat," *Ad Astra*, May 1991, 12.
90. Ibid.
91. Ibid., 11.
92. Ibid.
93. Ibid., 9.
94. The year 1995 has also turned out to be one of the hottest years on record.
95. National Aeronautics and Space Administration, "Leadership and America's Future in Space" (Washington, D.C.: NASA, August 1987)
96. Andrew Lawler, "Small Satellites Urged for Some Portions of EOS," *Space News*, 27 August-2 September 1990, 1, 20.
97. Andrew Lawler, "Earth Observation Plans Facing Changes," *Space News*, 16-22 September 1991, 1, 20.
98. Douglas Isbell, "EOS Platforms Will Shrink, Multiply to 18," *Space News*, 19-25 August 1991, 1.
99. Ibid.
100. Dave Cravotta, "Mission to Planet Earth: EOS-AM1," *Final Frontier*, April 1993, 34. Office of Technology Assessment, Solar Power Satellites, Washington, D.C.: GPO, 1981.
101. Ibid., 31.
102. "The Environmental Impact of SPS: A Social View," G.E. Canough and L.P. Lehman, Space Manufacturing 8: Energy and Materials from Space, (Washington, D.C.: American Institute of Aeronautics and Astronautics and Space Studies Institute, 1991), 34.
103. Ibid.
104. Ibid.
105. Ibid., 36.
106. Institute of Design, Illinois Institute of Technology, Systems and Systematic Design, Project Phoenix (Chicago: Illinois Institute of Technology, 1990), 1.
107. NASA's space station program has undergone many changes over the years and the current configuration is different from that of Freedom. The space station is now an international facility that primarily blends Russian and American space hardware. There will also be hardware contributed by Japan, the European Space Agency (ESA), and Canada. Construction is estimated to begin around 1997 or 1998.
108. International Space University, Summer Session 1992, Space Solar Power Program, (Kityakushu, Japan: International Space University, 1992), 12.

109. Ibid., 13.
110. Ibid., 14.
111. Ibid., 12.
112. Ibid., 16.
113. Ibid., 17-18.
114. Ibid., 15-16.
115. M.N.A. Peterson, David Criswell and Dan Greenwood, "Rationale and Plans for a Lunar Power System," Battelle Pacific Northwest Laboratory Power Beaming Workshop, May 14-16, 1991.
116. Ibid., 6.
117. Ibid.
118. David R. Criswell and Russell G. Thompson, "Decision Envelopment Analysis of Space and Terrestrially Based Large Scale Commercial Power Systems for Earth," 43rd Congress of the International Astronautical Federation, August 28-September 5, 1992.
119. M.N.A. Peterson, David Criswell and Dan Greenwood, "Rationale and Plans for a Lunar Power System," Battelle Pacific Northwest Laboratory Power Beaming Workshop, May 14-16, 1991, 9.
120. International Space University, Summer Session 1990, International Asteroid Mission, (Toronto: International Space University, 1990), xlix.
121. There were 76 shuttle flights through April 1996.
122. This is NASA press release 93-181, dated October 8, 1993.
123. Paul Hoverston and Steve Marshall, "Focus on Hubble's Vision," USA Today, 7 December 1993.
124. Dixon P. Otto, "On Orbit: Bringing on the Space Shuttle" (Athens, OH: Main Stage Publications, Inc., 1986), 4.
125. Ibid.
126. William Harwood, "Launch Cost of a Shuttle: Take Your Pick," Space News, 30 November-6 December 1994, 4.
127. Ibid., 29.
128. Ibid.
129. Ibid.
130. Liz Tucci, "Goldin: Determine Shuttle's Fate by 1997," Space News, 7-13 December 1992, 6.
131. U.S. National Aeronautics and Space Administration, 1995 Budget Summary.
132. Ibid.
133. Douglas Isbell and Lon Rains, "Leakage Trouble Possibly Unique to each Shuttle Orbiter," Space News, 16-22 July 1990, 18.
134. Douglas Isbell, "Shuttle Leak Streak Hits Fifth Month," Space News, 24-30 September 1990, 32.
135. National Commission on Space, Pioneering the Space Frontier (Toronto: Bantam Books, Inc., 1986), 12.
136. SDIO was renamed as the Ballistic Missile Defense Organization (BMDO) a few years ago.
137. William Boyer, "Report Proposes Fates for DC-X," Space News, 11-17 July 1994, 11.
138. Yankelovich, Skelly and White/Clancy Shulman, Public Opinion on the United States Civilian Space Program (Yankelovich, Skelly and White/Clancy Shulman, 1992), 1.
139. Ibid., 22.
140. Andrew Kohut and Larry Hugick, Gallup Poll: 20 Years After Apollo 11, Americans Question Space Program's Worth (Princeton: Gallup Poll News Service, 1989).
141. Ibid., 2.
142. Ibid.
143. George Gallup Jr. and Frank Newport, "Gallup Poll: NASA Rating Up, but Public Reluctant to Spend More Money on Space" (Princeton: Gallup Poll News Service, 1991).
144. Paul S. Hardersen, "The Challenger Learning Center in Des Moines opens the path to space," The Consortium Comet, August 1993, 1, 2.

Index

TO ORDER

additional copies of this book, other bestsellers in this *Frontiers in Astronomy and Earth Science* series, or other ATL Press professional science titles, please contact:

ATL PRESS, Inc., Science Publishers
Distribution Center, P.O. Box 4563 T Station, Shrewsbury, MA 01545, USA
ORDER TOLL FREE AND USE YOUR CREDIT CARD (all major cards)
Tel: 1-800-835-7543 (USA & Canada); 1-(508)-898-2290
FAX: 1-(508)-898-2063; E-Mail 104362,2523@compuserve.com

THE COMET HALE-BOPP BOOK

Guide to an Awe-Inspiring Visitor from Deep Space
Thomas Hockey
Foreword by comet co-discoverer Thomas Bopp
Frontiers in Astronomy and Earth Science, Volume 1

This book fills the long-standing demand for a comet guidebook for the curious 9th grader or 90-year old. Biographical details and quotes from Alan Hale and Thomas Bopp humanize the science behind the comet and capture the "feel" of what could well be this decade's astronomical event. The book provides charts and tables to locate and observe the comet. Hockey lucidly tells the story of the comet's discovery, astronomer's reaction to it, and their preparation for its arrival. He also covers other famous comets, such as Comet Halley, Comet Shoemaker-Levy 9 and Comet Hyakutake. Filled with color and black and white photos and illustrations, this book is an invaluable resource for all comet enthusiasts. Internationally-renown, Dr. Hockey is a professional planetary astronomer and experienced science educator.

Selected "Main Feature" of all Newbridge Science Book-of-the-Month Clubs

1996, 176 pages, illustrated, 6 x 9
ISBN 1-882360-14-1, 1996, Cloth US $42.90 (Outside US $52.50)
ISBN 1-882360-15-X, 1996, Paper US $19.95 (Outside US $25.50)

COMET HALE-BOPP T-SHIRT
COMET HALE-BOPP-THE COMET OF THE CENTURY T-SHIRT
THE YEAR OF THE COMET, 1996-1997 T-SHIRT

each $15.95 (100% cotton, assorted colors, sizes S, M, L, XL)

THE EARTH'S SHIFTING AXIS
Clues to Nature's Most Perplexing Mysteries
Mac B. Strain
Frontiers in Astronomy and Earth Science, Volume 2

What could trigger earthquakes and volcanoes, freeze mammoths, prompt the dinosaurs' disappearance, cause the climate changes prior to ice and coal ages, move tectonic plates, and create submarine canyons and fjords? After a lifetime of research and a prominent career at the U. S. Geological Survey, Strain presents new answers for many of Nature's mysteries. His revolutionary *Dynamic Axis Theory* holds our planet's nomadic North Pole responsible for many of these phenomena. Strain presents evidence for occasional earth axis surges and their impact on global evolution. Challenging current scientific thought, he sparks a fundamental debate about what makes our world function.

1996,appr. 250 pages, illustrated, indexed, 6 x 9
ISBN 1-882360-30-3, Cloth US $34.95 (Outside US $39.50)
ISBN 1-882360-31-1, Paper US $19.95 (Outside US $29.50)

THE CASE FOR SPACE
Who Benefits from Explorations of the Last Frontier?
Paul S. Hardersen
Frontiers in Astronomy and Earth Science, Volume 3

1996,194 pages, illustrated, indexed, 6 x 9
ISBN 1-882360-47-8, Cloth US $34.95 (Outside US $42.50)
ISBN 1-882360-48-6, Paper US $19.95 (Outside US $29.50)

Frontiers in Biomedicine and Biotechnology
VOLUME 1 CARBOHYDRATES AND CARBOHYDRATE POLYMERS
        M. Yalpani, Editor   ISBN 1-882360-40-0, 1993
VOLUME 2 LEVOGLUCOSENONE & LEVOGLUCOSANS
        Z. J. Witczak, Editor   ISBN 1-882360-13-3, 1994
VOLUME 3 BIOMEDICAL FUNCTIONS AND BIOTECHNOLOGY OF
        NATURAL AND ARTIFICIAL POLYMERS
        M. Yalpani, Editor   ISBN 1-882360-02-8, 1996

Frontiers in Foods and Food Ingredients
VOLUME 1 SCIENCE FOR THE FOOD INDUSTRY OF THE 21ST CENTURY
        M. Yalpani, Editor   ISBN 1-882360-45-1, 1993
VOLUME 2 NEW TECHNOLOGIES FOR HEALTHY FOODS &
        NUTRACEUTICALS
        M. Yalpani, Editor   ISBN 1-882360-10-9, 1996